REFLECTIONS ON SCIENTIFIC ATTITUDE

REFLECTIONS ON SCIENTIFIC ATTITUDE

Editor

Dr. DIGUMARTI BHASKARA RAO
M.Sc., M.A., M.A., M.Ed., Ph.D.

R.V.R. COLLEGE OF EDUCATION
GUNTUR—522 006
ANDHRA PRADESH
INDIA

DISCOVERY PUBLISHING HOUSE
NEW DELHI—110 002

First Published— 1997
Reprinted-2007

ISBN 81-7141-328-5

Published by :

Discovery Publishing House
4831/24, Ansari Road, Prahlad Street
Darya Ganj, New Delhi—110 002 (INDIA)
Phone : 327 92 45
Fax : 91-11-3253475

Laser Typesetting by :

Allied Computers,
Karnal (Haryana)

Printed at :
Sachin Printers, Delhi-

Preface

Scientific attitude is one of the key objectives of science teaching and it is also one of the major outcomes of it. Scientific attitude makes the people live as efficient citizens in the present scientific society. It also helps the people live upto their expectations and satisfactions. Understanding the role of scientific attitude in the life of a successful man, it is incorporated in all types of instructions and in all walks of education as a compulsory part, directly or indirectly. Many ways and means are used and applied to develop scientific attitude at various levels of education.

This book is aimed at providing the available material on scientific attitude alongwith some new thoughts. Since a long time, I have contacted many authors, researchers, educationists, editors and publishers working in India and abroad to collect the published and unpublished material on scientific attitude. Many gentlemen have extended their support either by writing a fresh paper or by giving permission to use their published works, orally or in writing. I am thankful to all of them who extended their cooperation. I also extend my apologies to the concerned authors and authorities wherever I failed to get the needful.

This book is a tribute to the authors and thinkers, who always advocated the culture of scientific attitude. I cordially invite papers on scientific attitude which will be incorporated in the forthcoming editions.

I feel it honoured if this book helps its readers to contribute for the cause of scientific attitude.

Dr. D. Bhaskara Rao
Editor,
Science Promoter
Vignan,
Guntur–522 002

Acknowledgements

I cordially thank
all the authors, editors and publishers
who contributed papers and
gave permission to use the published works
to compile this resource book on
scientific attitude

D. B. Rao
Sai Soudha
D-43, S.V.N. Colony,
Guntur–522 006
Andhra Pradesh
India

Contents

Preface V

Acknowledgements VII

1. Place of Scientific Attitude in the Objectives of Science Teaching
— Digumarti Bhaskara Rao 1

2. The Scientific Attitude and Science Education : A Critical Reappraisal
— Colin Gauld 8

3. What is the Scientific Attitude ?
— Robert L. Ebel 27

4. Scientific Attitudes
— Richard H. Lampkin, Jr. 44

5. Scientific Attitude
— M.S. Yadav 54

6. The Scientific Attitude
— Victor H. Noll 56

7. Scientific Attitudes
— R.C. Sharma 58

8. Scientific Attitude
— Otis W. Caldwell
Francis D. Curtis 60

9. Emphasise Scientific Attitudes in Teaching and Learning of Science
— Nathan S. Washton 62

Contents (Contd.)

10. Components of the Scientific Attitude
—Paul B. Diederich 63

11. Scientific Attitude
—Narendra Vaidya 68

12. Scientific Attitude
—M.S. Chhikara
S. Sharma 70

13. Scientific Attitude
—Edward W. Smith
Stanely W. Krouse, Jr.
Mark M. Atkinson 72

14. Scientific Attitude
—Pauline V. Young
Calvin F. Schmid 78

15. Scientific Attitude
—V.K. Kohli 83

16. Learning Scientific Attitudes
—Nelson B. Henry 90

17. Development of Scientific Attitude
—R.C. Das 97

18. Techniques for Developing Scientific Attitudes
—Eldwood D. Heiss
Ellsworth S. Obourn
Charles W. Hoffman 100

Contents (Contd.)

19. Development of Scientific Attitude
— J. K. Sood 108

20. Development of Scientific Attitudes
— R. Will Burnett 110

21. The Development of Scientific Attitudes
— Richard E. Haney 114

22. Developing Scientific Attitude
— Karuna Shankar Misra 123

23. Types of Learning Experiences in Science Helpful in Developing Attitudes
— Eldwood D. Heiss
Ellsworth S. Obourn
Charles W. Hoffman 128

24. Techniques for Developing Scientific Attitude
— K. Yadav 130

25. Acquisition of Certain Attitudes or Mind-sets
— Eldwood D. Heiss
Ellsworth S. Obourn
Charles W. Hoffman 135

26. The Role of Science Teacher in Developing Scientific Attitude
— Shaheen Usmani 138

27. An Attempt to Measure Scientific Attitudes
— Howard B. Baumel
J. Joel Berger 144

Contents (Contd.)

28. The Measurement of Scientific Attitudes
    — Ira C. Davis 149

29. An Approach to Measuring Scientific Attitudes
    — M. James Kozlow
    Marshall A. Nay 160

30. The Development and Application of a Scale for Measuring Scientific Attitudes
    — Victor Y. Billeh
    George A. Zakhariades 194

31. Scientific Attitude Scale
    — N. N. Shrivastava 209

32. Scientific Attitude Scale
    — J. K. Sood
    R.P. Sanadhya 219

33. Development of Scientific Attitude : An Experimental Study
    — M. J. Ravindranath 232

34. A Study of the Scientific Attitude and its Measurement
    — N. N. Shrivastava 239

35. Scientific Attitudes of In-service and Pre-service Science Teachers
    — D. Bhaskara Rao
    B. Joseph Raju
    G. Sundara Rao 242

Contents (Contd.)

36. Scientific Attitudes of Experienced Science Teachers
— Digumarti Bhaskara Rao
G. Sundara Rao
S. Raja Mohan Rao 250

37. Scientific Attitudes and Personality Traits of Prospective Science Teachers with High Pedagogic Aptitude
— D. Bhaskara Rao
G. Sundara Rao
S. Aruna
L. Rathaiah 256

38. Scientific Attitude in Secondary School Students
— Digumarti Bhaskara Rao 266

39. A Comparative Study of Scientific Attitude, Scientific Aptitude and Achievement in Biology at Secondary School Level
— Digumarti Bhaskara Rao 279

40. Brief Summary of Literature on Scientific Attitude
— Digumarthi Harshitha
Digumarti Pushpa Latha 296

1

Place of Scientific Attitude in the Objectives of Science Teaching

Digumarti Bhaskara Rao

Objectives in any area of curriculum should be regarded as direction of growth and not as ultimate ends to be completely reached. In this respect science is not different from other branches. It is important that objectives should be selected towards which the growth and development of the individual may be directed from a very practical point of view. Objectives need to be selected and stated in such a way that progress towards their attainment may be appraised.[1]

A judicious formulation and selection of worthwhile objectives for any school subjects goes a long way in enriching in shaping both the teaching and testing in that subject and such objectives should be evolved in relation to the needs of the individual in his society. The main sources for the formulation of the objectives are -(a) the needs and capabilities of the pupil, (2) the specific demands of his social environment (3) the nature of the subject matter, (4) the availability resources, and (5) the nature of the educational system.

Science teachers have long recognized the need for sound objectives in curriculum planning. In an examination of over 3,000 statements written from 1901 to 1950 by secondary school teachers, Paul Hurd[2] noted that, the objectives of science teaching were the teacher's first consideration in planning curriculum. Objectives strongly influence the organization of the curriculum and at the same time they provide the guide lines for the selection of teaching techniques.

National Society for the Study of Education in its Yearbook (1947) published the following objectives under categories, viz., (1) Functional information of habits, (2) Functional concepts, (3) Functional understanding of principles, (4) I skills, (5) Problem solving skills, (6) Attitudes, (7) Appreciations, and (8) Interests.

Bloom, *et al.*[3] classified the educational objectives into three domains, viz, the cognitive, the affective, and the psychomotor. The cognitive domain includes those objectives which deal with the recall or recognition of knowledge and the development of intellectual abilities and skills. The affective domain include objectives which describe changes in interest, attitudes, and values, and the development of appreciations and adequate adjustment. The work done in that period on manipulative or motor-skill was very less which includes motor activities.

Rai[4] in his Report on School Science Teaching states that the main objectives for teaching of science should be-

1. To arouse the curiosity of the student about the world we live in and to encourage him to understand the various natural phenomena.
2. To train to acquire the habit of making observation in a planned way.
3. To develop in him scientific attitude.
4. To give him an idea how a scientist works.

The aims and objectives of teaching general science, according to All India Seminar on Teaching of Science,[5] should be :

1. To familiarize the pupil with the world in which he lives and make him understand the impact of science so as to enable him to adjust himself to his environment.
2. To acquaint him with the scientific method and enable him to develop scientific attitude.
3. To give the pupil a historical perspective, so that he may understand the evolution of scientific development.

The Directorate of Extension Programmes for Secondary

Education, Government of India, in its brochure on 'Evaluation in General Science[6] sets some of the objective of teaching general science in school as follows :

1. The pupils studying general science should acquire knowledge of the fundamentals of science useful to all in everyday life.
2. They should develop the ability to apply the knowledge in everyday life.
3. They should acquire experimental skills such as : (a) handling apparatus and instruments ; (b) arranging apparatus for an experiment; and (c) preserving apparatus, chemicals, specimens, models, etc.
4. They should acquire constructional skills such as : (a) improvising simple instruments and appliances, and (b) repairing certain instruments and appliances of everyday life.
5. They should develop drawing skills such as :(a) drawing and sketching certain objects, instruments and arrangements ; and (b) photography in certain objects and specimens.
6. They should be able to locate reliable and recent information from appropriate sources.
7. They should be able to interpret scientific data given in various forms such as tabular, graphical, scientific etc.
8. They should develop the power of minute observation of their surroundings.
9. They should develop the power of oral expression in science to discuss, argue, describe and raise questions, using scientific terminology.
10. They should develop the scientific method in thinking and action.
11. They should adopt the scientific attitude in making statements, accepting information and forming beliefs.
12. They should develop interest in scientific reading and hobbies.
13. They should be able to appropriate the impact of science on life, both personal and social, the struggle through which science has advanced, and the inspiring work of the scientists.

The following similar set of objectives was formulated by the principals of Delhi Higher Secondary Schools[7] in the third summer camp organized by the Extension Department of the Central Institute of Education, Delhi.

1. To develop in the student a scientific attitude.
2. To develop in the student critical thinking.
3. To enable the student to acquire the fundamentals of scientific method.
4. To develop the student to be creative.
5. To develop in the student skill in laboratory techniques.
6. To develop in the student the ability to apply scientific knowledge and principles to problems of everyday life and new situations.
7. To enables the student to comprehend scientific terms, concepts, symbols, various tables and their uses.
8. To enable the student to construct and interpret graphs, diagrams and models.
9. To enable the student to collect and interpret data for the solution of problems.
10. To enable the student to be familiar with the natural resources of his environment and their uses.
11. To enable the student to be familiar with the trends in modern science.
12. To enable the student to appreciate the beauty and order in nature.

Approach Paper on Science and Mathematics in General Education[8] thought of the possibility of fulfilling the objectives for secondary level for enabling the students to :

a. study a few aspects of physical and life sciences in detail with special emphasis on those areas of concern like food, shelter, health, energy nutrition, and major components of environment;
b. appreciate the need of quantification in the scientific studies;
c. develop in science and ability to put the interest into action;
d. manipulate tools equipment in a proper manner;

e. identify the factors operating in the environment; and

f. collect data, classify and draw reasonable inference.

These above objectives of secondary stage are to be fulfilled along the objectives of primary and middle stages which include-collection of information ; classification of objects, events, etc.; identification of cause and effect relationship; development of scientific attitudes, acquainting with natural phenomena; giving emphasis to the relevance of science to daily life, etc.

Bhaskara Rao[9] states that a teacher must formulate some definite objectives and specification........ in order to achieve desirable behavioural changes among pupils. He emphasizes on objectives such as knowledge, understanding, application, skill, interest, scientific, attitude and appreciation.

First Asian Regional Conference on School Biology[10] held at Manila from December 4-10, 1966 recommended the following aims and objectives of school biology teaching in Asia.

1. To develop and instil in student the scientific attitude of inquiry and experimentations,
2. To provide sufficient understanding of the concepts of biology of enable students to become worthy citizens of the world.
3. To provide the opportunities for a practical understanding of the method of biologists which give them confidence to attempt the solution of problems which they have to face in their individual and social lives.
4. To give the student the incentive to pursue the study at higher levels of biology and related fields.
5. To encourage respect and feeling for living things.

National Council of Educational Research and Training[11] put for the following objectives for biological science.

1. The pupil acquires knowledge of biological terms, concepts, principles, formulae, etc.
2. The pupil understand biological terms, facts, concepts, principles and processes, etc.
3. The pupil applies knowledge of biology in new situations.

4. The pupil develops skills in

   (a) drawing, (b) manipulating, (c) collecting, (c) dissecting, (e) observing biological specimens, and (f) locating biological information.

5. The pupil develops interest in the living world.

6. The pupil develops scientific attitude towards scientific phenomena.

7. The pupil appreciates the contribution of science to human welfare.

And the latest National Policy on Education - 1986[12] states that 'Science Education will be strengthened so as to develop in the child well defined abilities and values such as the spirit of inquiry, creativity, objectivity, the courage to question, and an aesthetic sensibility.'

All the above aims and objectives of science stress, directly or indirectly, the importance of scientific attitude along with scientific aptitude, skills, abilities and interests.

REFERENCES

1. Eldwood D. Heiss, Ellswroth S. Oborun and Charles W. Hoffman, Hoffman, *Modern Science Teaching*, (New York : The Macmillan Co., 1950), p. 24-25.
2. Hurd, P.H., "The Education Concepts of the secondary Science Teaching", *School Science and Mathematics* (1954), 89-96.
3. Benjamin S. Bloom, ed., *Taxonomy of Educational Objectives, Handbook 1 : Cognitive Domain* (New York : Longman, Green and Co. 1956), p.7
4. R.N. Rai, *Report on School Science Teaching*, Seminar held at the University of Cylon, December 1963. As cited by D. Gopala Krishna, *A survey of Scientific Attitude and Relation to Intelligence of Graduate Students* (Unpublished M.Ed. Dissertation, Andhra University, Waltair, 1975), p.10
5. Proceeding of the All India Seminar on the Teaching of Science in Secondary Schools, 1956.

6. R.C. Das, *Science Teaching in Schools* (New Delhi ; Sterling Publishers Private Limited), pp. 22-23.
7. Ibid., p.23.
8. *Approach Paper on Science & Mathematics in General Education*, Report of the working group on Science & Mathematics, Sept. 1985 (Dept. of Education in Science & Mathematics, N.C.E.R.T., New Delhi), pp. 5-6.
9. D. Bhaskara Rao, "Objectives of Science", *Science Promoter*, 2 (October 1989), 701-703.
10. *First Asian Regional Conference on School Biology*, As cited by M.S. Chhikara and S. Sharma, *Teaching of Biology* (Ludhiana: Prakash Brothers, 1985), pp. 19-20.
11. Pritam Singh, *A Monograph on Improving Practical Examination in Science* (New Delhi : N.C.E.R.T., 1983) pp. 15-19.
12. *National Policy on Education - 1986*, Department of Education, Ministry of Human Resources Development, Government of India, New Delhi, May 1986, p. 23.

2
The Scientific Attitude and Science Education : A Critical Reappraisal

Colin Gauld

The Nature of the Scientific Attitude in Science Education

For more than 60 years science educators have included the development of the scientific attitude among the general aims of science education. Some writers label this attitude as "scientific-mindedness" (Burnett, 1944), "the habit of scientific thinking" (Noll, 1933a) or "the spirit of science" (Educational Policies Commission, 1966) and it is most often characterized by a list of component attitudes ("scientific attitudes" such as objectivity, open-mindedness, scepticism, and a willingness to suspend judgment if there is insufficient evidence.

Many writers have pointed out that knowledge about scientific facts and skill in the use of scientific methods are of little value if there is no inclination to use them. The scientific attitude represents the motivation which converts this knowledge and skill into action and refers to a *willingness* to use scientific procedures and methods. It may best be described as "an attitude to ideas and information to particular ways of evaluating them", a formulation which distinguishes it from "an attitude to science or scientists" on the one hand and from "an ability to carry out scientific procedures" on the other (Gauld & Hukins, 1980).

Major statements of the goals of science education in the USA have consistently stressed the importance of developing scientific attitudes in students (see, for example, Whipple, 1932; Henry, 1947; Henry, 1960; Educational Policies Commission, 1966; N.S.T.A., 1971) and curriculum projects around the world include this among their aims (Gauld & Hukins, 1980). In an analysis of 1,547 aims culled from the science education literature Fraser (1977) found that almost half of these could be categorized as aims related to the development of the scientific attitude. However, there is a deal of evidence that little emphasis is placed on this aim in the classroom, apparently because methods for teaching and testing attitudes may not be widely available rather than because of a general dissatisfaction with the aim itself.

The scientific attitude as it appears in the science education literature embodies the adoption of a particular approach to solving problems, to assessing ideas and information or to making decisions. Using this approach evidence is collected and evaluated objectively so that the idiosyncratic prejudices of the one making the judgment do not intrude. No source of relevant information is rejected before it is fully evaluated and all available evidence is carefully weighed before the decision is made. If the evidence is considered to be insufficient then judgment is suspended until there is enough information to enable a decision to be made. No idea, conclusion, decision or solution is accepted just because a particular person makes a claim but it is treated sceptically and critically until its soundness can be judged according to the weight of evidence which is relevant to it. A person who is willing to follow such a procedure (and who regularly does so) it said by science educators to be motivated by the scientific attitude.

It is clear that, in the minds of many writes, "evidence" means, "empirical evidence" (Downing, 1928; No., 1933a; Ward, 1933; Henry, 1947, pp. 168–171; Lampkin, 1951; Educational Policies Commission, 1966, p. 19; Diederich, 1967; Collette, 1973, pp. 14-15, 20; Sund and Trowbridge, 1973, pp. 5–7) and the discussion usually implies that empirical evidence is the only type of evidence which needs to be considered in making scientific decisions. The ultimate test in science is how the conclusion fits with the facts. Thus a person who is motivated by the scientific attitude as it is generally conceived by science educators is someone who makes decision solely on the basis of the weight of empirical evidence and this view of the scientific attitude will be labelled here as "empiricist". Lampkin (1951), Feigl (1955) and Kurtz (1976) have clearly shown how such a conception is closely related to a particular view of a knowledge in general and of science in particular—a philosophical perspective which has also been labelled "empiricist".

The empiricist attitude in science education takes one of two forms depending on the type of decision which is presumed to follow from a consideration of the empirical evidence relevant to a theory. In the "verificationist" version, empirical evidence is used to *verify* or *prove* the truth of a proposition or hypothesis (Downing, 1928; Ward, 1933; Lampkin, 1951; Van Deventer, 1960, p 104; Diederich, 1967). Diedrich (1967) includes "a desire for experimental verification" as a component of the scientific attitude, while Lampkin (1951) defines the component labelled "scepticism" as "an unwillingness to accept statements which are not supported by evidence defined as verification of predictions". Ward (1933) sees science as "the body of experience and theory that can be verified by all observed alike" and, for Kurz (1976) "a belief is true if, and only if, it has been confirmed, directly or indirectly, by reference to observable evidence."

However, the realization that logically the truth of a universal proposition cannot be finally proven by appealing to a finite number of items of confirming data has led some writers to adopt a "falsificationist" version of the empiricist attitude. Empirical evidence allows one to state unambiguously, not when a theory is true, but when it is false.

A single new scientific fact disagreeing with the theory completely invalidates the theory. The willingness to give up an old established theory as soon as it is proved to be definitely inconsistent with a single fact is the attitude of Science; no branch of knowledge without this attitude can be called a science (Podolsky, 1965).

Reasons for the Development of the Scientific Attitude of Students

For many science educators the importance of the scientific attitude is so obvious that no argument is required to support its inclusion among those which a school science course should aim to develop in students. This idea is also reinforced by the fact that there has been little, if any, argument *against* its inclusion among the aims of science education. However, while it may be obvious that the scientific attitude is important in the professional lives of scientists and that students learning about science should also become aware of the motive power which impels scientists in their work, it is not a simple matter to move on to the conclusion that school students, many of whom do not intend to become scientists, should actually be encouraged to adopt this attitude for themselves.

Two types of argument are offered by those who do provide reasons for taking this final step. In the first, it is argued that an effective way of learning about the nature of scientific activity is for the student to act out the role of a scientist in the classroom.

> Every child—not just those who manifest interest or high motivation—must be viewed as a young scientist by the teacher of science . . . . . (he or she) must experience the mode and the excitement and the frustration of the scientist. (Link, 1967).

The student who enters this role fully will be the one who adopts for himself the attitude which also motivates the scientist (Nay and Crocker, 1970).

In the second type of justification it is argued that not only does the adoption of the scientific attitude for themselves help students to understand the nature of science and the activities of scientists better but scientific attitudes represent desirable personal attributes for all people. The tendency to be accurate, intellectually honest open-minded, objective, and to demand reliable empirical evidence before making decisions may be most clearly seen in the problem solving activity of scientists (so this arguments goes) but they also represent predispositions appropriate for solving problems in everyday life as well. Under the influence of such attitude as these, it is claimed that problems will be approached in a manner which is more likely to lead to successful solutions (see, for example, Noll 1933b). For the Educational Policies Commission (1966) possession of the scientific attitude is not only the mark of a scientifically— minded person, but also the sign of a rational one. These benefits of a scientific education are primarily for the individual but a number of writers have claimed additional benefits for the society.

> As we consider the future responsibilities of citizens we will probably agree that helping children to become more co-operative, more responsible, more 'open-minded', and, at the same time, more 'critical-minded' is certainly worth the effort. (Henry, 1947, p. 87).

By adopting scientific attitudes and transferring these to situations in everyday life, students, can be expected to more tolerant of other points of view and to be more successful in living and working alongside other people. According to the Educational Policies Commission science can provide "power, prestige, standard of living, education, and health" but the spirit or science" promises two less tangible but equally profound benefits: increased individuality and increased brotherhood of men" (1966, p.11).

Behind both these arguments is the assumption that scientists really are motivated by the scientific attitude as it is presented by science educators (Offner, 1937, Haney, 1964; Diederich, 1967). In other words, in solving scientific problems, scientists adopt an empiricist attitude in which empirical data, gathered objectively is the final judge of truth, if not in accepting as true those hypotheses which are supported by the evidence, then certainly in rejecting as false those which conflict with it; an attitude in which the ideas of other scientists are received in an open-minded manner and given full, impartial but critical consideration.

Scientists and the Scientific Attitude: Research Evidence

In science education over the past sixty years a great deal of effort has been devoted to identifying the nature of the scientific attitude and most of this work has been based on detailed analyses of the writings of scientists, philosophers of science, and science educators (Curtis, 1926; Noll, 1933a; Davis, 1935; Crowell, 1937; Ebel, 1938; Lampkin, 1938; Vitrogan, 1967, 1969; Nay and Crocker, 1970; Cohen, 1971). In a number of cases, the results of these analyses were submitted to panels of scientists or science teachers to obtain estimates of the relative value of each component of the scientific attitude arising from the analyses. The primary source material for these investigations was the writings of philosophers of science who looked at science from an empiricist perspective and it is easy to understand why the conception of the scientific attitude which emerged also possessed an empiricist emphasis. It is interesting to observe that, in spite of the obvious value placed on empirical evidence by science educators who write about the scientific attitude, almost no interest has been shown in whether scientists do, in fact, possess the affective characteristics attributed to them on the basis of such analysis of the literature. It is difficult to find any reference in science education literature to studies of the psychology of scientists, of sociological research into the nature of the ethos of science, of recent historical case studies of the activities of scientists, or of alternatives to the empiricist model of science which seem to lie behind the science educator's conception of the scientific attitude. For the past thirty years, relevant information from the these areas has been accumulating and will be reviewed in the following sections.

The Psychology of the Scientist

In the early 1950's Roe carried out extensive psychological studies of eminent physical and biological scientists, anthropologists and psychologists. She reported some of her conclusions from these investigations in the following way:

> Characterizations of scientists almost always emphasize the objectivity of their work and describe their cold, detached, impassive, unconcerned observation of the phenomena which have no emotional meaning for them. This could hardly be further from the truth . . . . the creative scientist whatever his field, is very deeply involved emotionally and personality in his work . . . I think many scientists are genuinely unaware of the extent, or even of the fact, of this personal involvement, and themselves accept the myth of impersonal objectivity (Roe, 1961).

Eidusion's study of the psychological world of the scientist also led her to conclude that scientists themselves misrepresent the diversity of the personal characteristics which exist within their occupational group and so "perpetuate some of the fixed and stereotyped notions that exists about a scientist" (1962, p.250; see also pp. 124, 153, 154, 255).

The more recent work of Mahoney (1976, 1979) in this area was directed towards examining the extent to which scientists possess the characteristics— objectivity rationality, open-mindedness, superior intelligence, integrity, and communality—that the writings of scientists and science educators attributed to them. His description of the "real" scientist departs considerably from the picture presented in the science education literature. He arrived at the following conclusions;

1. Superior intelligence is neither a prerequisite nor a correlate of scientific contribution;
2. The scientist is often saliently illogical in his work, particularly when he is defending a preferred view or attacking a rival one;

3. In his experimental research, he is often selective, expedient, and not immune to distorting the data;
4. The scientist is probably the most passionate of professionals; his theoretical and personal biases often color his alleged "openness" to the data.
5. He is often dogmatically tenacious in his opinions, even when the contrary evidence is overwhelming.
6. He is not the paragon of humility or disinterest but is, instead, often a selfish, ambitious and petulant defender of personal recognition and territoriality;
7. The scientist often behaves in ways which are diametrically opposite to communal sharing of knowledge— he is frequently secretive and occasionally suppresses data for personal reasons; and
8. Far from being a "suspender of judgment" the scientist is often an impetuous truth spinner who rushes to hypotheses and theories long before the data would warrant (Mahoney, 1976, p. 6).

Mahoney's portrait of the "real" scientist is of someone of who displays both objectivity *and* emotionality, open-mindedness *and* tenacity, depending on the context. He speculates (and the truth of his speculation is borne out by the work of others) that it is the less eminent scientist who comes closest to possessing the qualities of the empiricist ideal (1979). Among a sample of forty-two scientists who were actively involved in research related to data obtained through the Apollo moon missions, Mitroff and Mason (1974) found a similar range of personality traits as that referred to by other investigators.

The single dimension which most served to differentiate between the scientists was that of "speculativeness" or "willingness to extrapolate beyond the available data". At one end of the spectrum where the extreme speculative scientists who in the words of the respondents "wouldn't hesitate to build a whole theory of the solar system based on no data at all"; on the other extreme were the data-bound, scientists who "wouldn't be able to save their own hide if a fire was burning next to them because they'd never have enough data to prove the fire was really there". On every subsequent dimension on which these two types of scientist were compared they stood in extreme contrast to one another. One of the most significant things about these differences is that the more outstanding a scientist was, a judged by his peers, the more he lay near the speculative end of the scale. Conversely, the more "mundane", "typical", or "run-of-the-mill" scientists fell toward the "data-bound" end of the scale......

At the same time the more speculative scientists are also the kinds of scientists who are more likely to become rigidly committed to their ideas once they have produced them. Contrary to popular misconception, it is the "lesser" not the "greater" scientist, who is more likely to have an "open mind". The greater the scientist the more likely he is to develop a line and to push it for all it is worth .... In a word, the greater the scientist, the more likely he is to believe the myth of the disinterested, uncommitted, scientist, (1974; see also Hilli, 1974).

There is one notable discrepancy between the observation of Roe and Eiduson on the one hand Mitroff on the other. While the former investigators found that the scientists they interviewed accepted the empiricist stereotype, "every one of the scientists interviewed on [Mitroff's] first round of interviewes indicated that they thought the notion of the objective, emotionally disinterested scientist naive" (Mitroff, 1974b). It may be that, when discussing the work they actually do, scientists are more aware of the inappropriateness of the empiricist stereotype than when they are talking about science in general (see Mitroff, 1974a, pp. 107-131).

The Ethos of science

In an article on the ethos of science, originally published in 1942, Metron described the ethos of science as "that emotionally toned complex of values and norms which is held to be binding on the man of science." Control over the scientist's behavior is imposed through these norms by sanctions and rewards and "are in varying degrees internalized by the scientist" (Metron, 1968, p. 605).

Merton identified "universalism", "organized scepticism", "communism", and "disinterestedness" as four norms through which he claimed institutional control was exerted over the behavior of scientists. To these Barber (1952, pp. 84-94) added "rationality" and "emotional neutrality" and Storer described the six norms as follows:

Universalism: This norm ... refers both to the assumption that physical laws are everywhere the same and to the principle that the truth and value a scientific statement is independent of the characteristics of its author . . .

Organized Scepticism: This norm (embodies) the principle that each scientist should be held individually responsible for making sure that previous research by others on which he bases his work is valid . . .

Communism, or Communality: This norms directs the scientist to findings with other scientist freely and without favor . . .

Disinterestedness: This norm ...... makes it illicit for the scientist to profit personally in any way from his research . . .

Rationality: (This is) a faith in the moral virtue of reason . . It may be interpreted also as the assumption that necessary to the achievement of the goals of science are (1) empirical test rather than tradition (2) a critical approach to all empirical phenomena rather than acceptance of certain phenomena as exempt from scrutiny . . . . . .

Emotional Neutrality: (This norm) enjoins the scientist to avoid so much emotional involvement in his work that he cannot adopt a new approach or reject an old answer when his findings suggest that this is necessary, or that he intentionally distorts his findings in order to support a particular hypothesis (1966, pp. 78-80).

Storer adds that "it is relatively easy to show that this combination of norms is admirably suited to ensure the optimal progress of science; ideally, only when scientists' behavior is guided by these norms is it possible to keep scientist in touch with the frontiers of knowledge" (1966, pp. 82,83). An examination of Storer's description of the norms of science show that it is an expression in sociological terms of the empiricist conception of the scientific attitude found in science education.

Later research by Merton (1963,1969) suggested that in addition to working under the control of norms such as those above, the scientist seemed also to be influenced by of set of what have been called "counter-norms". These represent pressure from the scientific institution to act legitimately (that is, in the interests of science) in the *opposite* direction to that specified by the original norms (see also Rothman, 1972).

One purpose of Mitroff's study of moon scientists was to investigate the extent to which norms and their counter-norms exercised control over scientists in their professional work. About forty eminent scientists who were directly interested in moon rock samples collected on the Apollo missions were interviewed four times over a span of 3½ years between appollo 11 and Appollo, 16. In addition, they are asked to respond to a number of tests and questionnaires. Extensive data were gathered which demonstrated the operation of both conventional norms and counter-norms with this groups of scientists and Mitroff produced an expanded list a basis for further study (1974a, p. 79; 1974b).

This lack of acceptance of the simple empiricist stereotype was found in an earlier study in which West (1960) carried out a survey of the scientific values of fifty-seven academic scientists at a Midwestern university. He found that there was a wide variation in the strength of adherence to "the classical ideology of science" and that there was little relationship between this and the extent of productive research.

Mulkay (1976, 1979) has argued that neither the norms nor the counter-norms referred to above are the strong determiners of behavior that Merton and Mitroff consider them to be, since it is not private behavior but the public presentation of results which determines the allocation of rewards in science. For him the so-called norms and counter norms constitute an informal, moral vocabulary "which scientists can use flexibly to categorise professional actions differently in various social contexts and, presumably in accordance with varying social interests" (1976; see also Barnes and Dolby, 1970). This vocabulary points specifically to problem areas in scientific practice including those related to objectivity-subjectivity, rationality-irrationality, and impartiality-commitment, but it contains no solutions to these problems. For example, a focus on the need for objectivity may be used to counteract an opponent who is judged to have been unduly subjective, while an appeal to the subjective aspects of science may be used where claims for objectivity appear to be excessive.

Historical Case Studies

Information about the scientific attitude is often conveyed and reinforced in an educational setting by appealing to the work of scientists in the past. In particular, support for the empiricist conception of this attitude is derived from the way scientists in the past apparently constructed theories and made decisions about their validity solely on the basis of experiments. It is claimed, for example, that the Michelson-Morly experiment simultaneously dealt the death blow to the Lorentz electron theory and led to the birth of Einstein's special theory of relativity; or that Millikan's oil-drop experiment once and for all settled dispute about the indivisibility of the electron charge.

In his history of the Michelson-Morley-Miller experiments before and after 1905, Swenson (1970) has shown that, following the 1887 version of the experiment, a number of explanations for the null result were still available which did not require the rejection of the concept of the aether. Because of this fact, modifications, of the original experiment were carried out over the next forty-five or fifty years to attempt to demonstrate conclusively the presence or absence of aether drift. In 1925, Miller announced that he had obtained aether drift velocity of about 200 meters per second but, instead of causing Einstein's theory to be rejected, this result was effectively ignored for the next thirty years, in spite of Miller's acknowledged competence. In 1954, Shankland suggested that the result may have been due to lack of adequate temperature control.

Holton's research (1969), and the work of others, has also shown that the Michelson-Morley experiment had little or no effect on the origin of the theory of relativity in Einstein's mind. Einstein was apparently prompted more by "the essential requirement of finding symmetry and universality in the operations of nature" (Holton, 1969) that by substantially empirical considerations. The myth which describes a direct link between the 1887 version of the Michelson-Morley experiment and Einstein's conception of the theory of relativity is, for Holton, an example of the effect of *experimenticism* which "is best recognized by the unquestioned priority assigned to experiments and experimental data in the analysis of how scientists do their own work and how their work is incorporated into the public enterprise of science." (1969).

The later study of Holton of Millikan's laboratory notebooks for the years 1911 and 1912, illuminates another aspect of the "experimenticist" notion of science. The results published in Millikan's 1913 paper so clearly support Millikan's view of the indivisibility of the electron charge that, at least in this case, it seems obvious that the oil-drop experiment conclusively decided the point the issue between Millikan and Ehrenhaft. Millikan announced there that "the largest departure from the mean value found anywhere in the table [of values of e, determined for fifty-eight droplets] amounts to 0.5 percent" (quoted in Holton, 1978, p. 61). However, Millikan's notebooks show that results for many more droplets were eliminated even as they were being obtained. Millikan "evaluated his data and assigned qualitative indications on their prospective use, guided by both a theory about the nature of the electric charge and a sense of the quality or weight of the particular run " (p.70). In order to account for Millikans' behavior in the laboratory, Holton introduces in notion of *suspension of disbelief* to describe the procedure of holding in abeyance "final judgements concerning the validity of apparent falsifications of a promising hypothesis" (p. 71) especially during the early stages of theory construction or testing. It may also be possible to explain inconsistencies in Mendel's published work in a similar way (Fisher, 1936).

Millikans' example demonstrates one possible response to data which conflict with expectations at the stage of scientific work prior to publication. However, even after results are published, a scientist whose theory is apparently falsified by the evidence can handle the situation in a number of different ways in order to retain his theory. He can deny the validity of the data and suggest possible reasons why it should be ignored; he can accept the data but give reasons why it has no serious implications for the theory; or he can accept both the data and the implications for the theory but argue that when all the problems have eventually been cleared up the theory will be vindicated (Mahoney, 1976, 1979; Popper, 1968, p. 50, Kugn, 1970). The treatment of Miller's 1925 results for the aether drift velocity is one example of the third approach to data which apparently falsifies a theory. In another example, clear deviations in the orbit of the planet Uranus from that expected on the basis of Newton's theory of gravitation did not lead to the rejection of theory. Instead they were eventually found to be caused by an originally unexpected factor—the existence of the planet Neptune—which had no part to play in the theory itself. For many years, similar deviations in the orbit of mercury were not counted against the theory because it was felt that they would eventually be explained when a further, as yet unobserved planet was found. Barber's survey (1961) of the extent of which scientists resist the introduction of new ideas also demonstrates the strength of opposition to falsifying evidence.

If as, the above (and many other) examples demonstrate, experimental evidence does not conclusively speak for or against a particular theory one is led to adopt a non-empiricist position similar to that of Einstein who, "though unwilling to accept the possibility of confirmation of a theory by "verification" of its prediction, ...in practice also held to the falsification principle only sceptically (weakly) when the theory purportedly falsified by experimental test had in his views certain other merits compared with its rivals" (Holton, 1978, p. 98).

Alternative Models of Science

The above evidence concerning the behavior of scientists and, in particular, their use of experimental data in coming to conclusions about theories is at odds both with the conception of the scientific attitude possessed by science educators and with an empiricist philosophy of science to which this attitude seems to be related (Feigl, 1955; Kurtz, 1976). If the "orthodox" view of scientific attitude is to be modified in a way which will accommodate the evidence presented above, it will also be necessary to change the model of science apparently adopted by science educators working in this area. Since about 1960, an increasing range of non empiricist philosophies of science ha become available from Kuhn's model of paradigm conflict (1962) to the anarchistic view of Feyerbend (1970). There is also evidence that strictly empiricist philosophies have been modified in the light of the mounting evidence against them (see, for example Scheffler, 1967).

These developments in the philosophy of science are well known and will not be further discussed here. Howsoever, both Holton and Mitroff, whose research has been presented above, have offered further suggestions about views of science which they consider to be consistent with the result of their work.

In 1952, Holton introduced the distinction between "public and "private" science (1952, pp. 234-256). The way in which arguments and evidence are publicly presented ("public" science) and not the way in which they were originally conceived, clarified, and tested ("private" science) is used by others to judge the scientific value of the work of a scientist. The technical format of a scientific paper is such that references to personal characteristics of the author are rigorously excluded. Holton suggested that the empiricist stereotype of the scientist as detached and impartial is one which arises from this edited, public image and not from a study of scientists themselves as they engage in "private" science. His own work in the history of science has provided considerable evidence that, in the private work of a scientist, the range of appropriate personal characteristics is almost unlimited. It certainly includes those which have been incorporated into the empiricist scientific attitude together with their opposites as outlined by Mitroff (1974a, p. 79; 1974b). If the distinction between "public" and "private" science is a valid one, it means that the attitudes toward scientists held by science educators and science students can be expected to have little, if any, necessary connection with the personal characteristics of scientists.

Mitroff's research raised for him the problem that if, as seems to be the case, the best scientists are those who tenaciously hold on to their theories almost in spite of the evidence them, how can science be considered to be objective even in the public domain? In order to solve this problem, he appealed to a dialectical or adversary notion of science (1972; 1974a, pp. 219-250). In his view, a theory a most likely to get a fair hearing if there are individuals passionately committed to its validity and who do all they can to produce evidence and arguments in its favour. Those most likely to come up with evidence and arguments against the theory are those who disagree with it or who are committed to an alternative view. The interactions between these two groups of passionately committed individuals helps to bring to light as much relevant evidence as possible for assessment by the relevant community of scientists without it being necessary for either group to show complete objectivity, open-mindedness, impartiality, or emotional neutrality. The similarity between this way of viewing science and the way truth is sought in a courtroom is obvious as are the parallels with Kugn's notion of paradigm conflict (1962). Of course the degree of commitment which scientist adopt towards a particular theory varies from individual to individual and the situation is not so clearly defined as may be suggested by the above discussion. But, while the empiricist stereotype allows for little variation among the personal attributes of scientists, the views of Holton and Mitroff appear to make more sense of their actual behavior.

Conclusions

Arguments for including the development of the scientific attitude in students among the main goals of science education rest firmly on the assumption that this attitude is demonstrated in the professional behavior of successful scientists. The conception of the scientific attitude which appears in the science education literature sees the scientist as some one who make decisions solely on the basis of empirical evidence and who at all times prevents his personal interests from intruding into these decisions. The evidence presented demonstrates clearly that this view, which seems to have been derived primarily from the writings of scientists and philosophers of science before about 1960, is completely untenable and may, at best, be associated with the less successful scientist.

The lack of attention which has been given by the educators who carry out research into the scientific attitude to the evidence presented here from the psychology, sociology, history and philosophy of science, may be just another example of commitment to one view (in this case, an empiricist view of science) can lead one to ignore contrary evidence. The proliferation, since 1960, of non empiricist philosophies of science, has had little obvious influence on how the scientific attitude is conceived by science educators. Even Klopfer's recent extensive and detailed outline of the structure of the affective domain in relation to science education (1976) retains many features of the empiricist stereotype with little acknowledgment that contrary evidence has been taken into account. One purpose of this paper has been to present the evidence against the empiricist conception of the scientific attitude.

A conclusion that could be drawn from the material and arguments presented here is that development of the scientific attitude in students should be eliminated as one of the major goals of science education, and this certainly follows for the attitude as it has been formulated by science educators for the past 60 years. Teaching that scientists possess these characteristics is bad enough but it is abhorrent that science educators should actually attempt to mold children in the same false image. On the other hand, very few writers explain what they mean by open-mindedness, objectivity, or scepticism, and little indication is given of how evidence is weighted or of how one decides when there is sufficient evidence to make a decision. It is possible that, if such terms were clarified and the way in which they relate to scientific practice were more carefully discussed in the light of the material presented here, one could retain a reformulated and more acceptable version of the scientific attitude.

This second alternative requires considerably more discussion that has taken place up to the present. If any conception of the scientific attitude is to be retained in science education it is no longer sufficient to build unquestioningly on the consensus of science educators. Too much relevant information has been ignored. If further work takes place which clarifies, in the light of the above evidence, the role which the development of the scientific attitude should play in science education then the second purpose of this paper will have been achieved.

REFERENCES

Barber, B. *Science and the Social Order*. Glencoe, III; Free Press, 1952.

Barber, B. Resistance by scientist to scientific discovery. *Science*, 1961, *134*, 596-602.

Barnes, S.B., & Dolby, R.G.A. The scientific ethos: A deviant viewpoint. *Archives Europeen de Sociologie*, 1970, 9, 3-25.

Burnett, R. W. The science teacher and his objectives. *Teachers College Record*, 1944, 45, 241-251.

Cohen, D. Can scientific attitudes be evaluated ? In *Research 1971*, R.P. Tisher (Ed.). Melbourne: Australian Science Education Research Association, 1971, 135-143.

Collete, A.T. *Science teaching in the secondary school*. Boston: Allyn and Bacon, 1973.

Crowell, V. The scientific method. *School Sci. Math.*, 1937, 37, 525-531.

Curtis, F.D. A determination of the scientific attitudes. *J. Chem. Educ.*, 1926, 3, 920-927.

Davis, I.C. The measurement of scientific attitudes. *Sci, Educ.*, 1935, 19, 117, 122.

Diederich, P.B. Components of the scientific attitude. *The Science Teacher*, 1967, 34(2), 23-24.

Downing, E.R. The elements and safeguards of scientific thinking. *Scientific Monthly*, 1928, 26, 231-243.

Ebel, R.L. What is the scientific attitude? *Sci. Educ.*, 1938, 22, 1-5, 75-81.

Educational Policies Commission, *Education and Spirit of Science*. Washington, D.C.: National Education Association, 1966.

Eiduson, B.T. *Scientists: their psychological world*. New York: Basic Books, 1962.

Feigl, H. Aims of education for our age of science: reflections of logical empiricist. In *The fifty-fourth yearbook of the National Society for the Study of Education, Part I: Modern philosophies and education*, N. B. Henry (Ed.) Chicago: National Society for the Study of Education, 1955, 304-341.

Feyerabend, P.K. Against method. *Minnesota Studies in the Philosophy of Science*, 1970, 4, 17-130.

Fisher, R.A. Has Mendel's work been rediscovered? *Annals of Science*, 1936, 1(2), 115-137, Fisher, B. Selection and validation of attitude scales for curriculum evaluation. *Sci, Educ.*, 1977, 61(3), 317- 330.

Gauld, C. F., & Hukins, A.A. Scientific attitudes: a review. *Studies in Science Education*, 1980, 7, 129-161.

Haney, R.E. The development of scientific attitudes. *The Science Teacher*, 1964, *31*(12), 33-35.

Henry, N. B. (Ed.) *The forty-sixth yearbook of the National society for the Study of Education Part I: Science Education in American Schools*, Chicago: National Society for the study of Education, 1947.

Henry, N.B. (Ed.) *The fifty-ninth yearbook of the National Society for the Study of Education. Part I: Rethinking science education.* Chicago: National Society for the study of Education, 1960.

Hill, S.C. Questioning the influence of a 'Social System of Science': a study of Australian scientist. *Science Studies*, 1974, 4, 135-163.

Holton, G. *Introduction to concepts and theories in physical science*, Reading, Mass: Addison-Wesley, 1952.

Holton, G. Einstein, Michelson and the "crucial" experiment. Issi, 1969, 60, 133-197.

Holton, G. *The scientific imagination: case studies.* Cambridge: Cambridge University Press, 1978.

Klopfer, L.E. A. structure for the affective domain in relation to science education. *Sci. Educ.*, 1976, *60*(3), 299-312.

Kuhn, T.S. *The structure of scientific revolutions.* Chicago: University of Chicago Press, 1962.

Kuhn, T.S. Logic of discovery or psychology of research ? In *Criticism and the growth of knowledge*, I Lakatos & A. Musgrave (Eds.) London: Cambridge University Press, 1970, 1-23.

Kurtz, P. The scientific attitude vs anti-science and pseudoscience. *The Humanist*, 1976, *36*, 27-31.

Lampkin, R.S. Scientific attitudes, *Sci. Educ.*, 1938, 22, 353-353.

Lampkin, R.S Scientific enquiry for science teachers. *Sci. Educ.* 1951, 35, 17-19.

Link F. An approach to a more adequate system of evaluation in science. *The Science Teacher*, 1967, *34*(2), 20-24.

Mahoney, M.J. *Scientist as subject: the psychological imperative.* Cambridge, Mass.: Ballinger, 1976.

Mahoney, M.J. Psychology of the scientist : an evaluative review. *Social Studies of Science*, 1979, *9*, 349-375.

Merton, R.K. The ambivalence of scientists. *The Bulletin of the Jogn Hopkins Hospital*, 1963, *112*,77-97.

Meton R.K. *Social theory and social structure*. New York: Free Press, 1968.

Merton R.K. Behavior patterns of scientists. *American Scientist*, 1969, 57(1), 1-23.

Mitroff, I. I. The myth of objectivity or why science needs a new psychology of science. *Management Science: Application*, 1972, 18, B613-B618.

Mitroff, I.I. *The subjective side of science*. Amersterdam: Elsevier, 1974a.

Mitroff, I.I. Norms and counter-norms in a select group of the Apollo moon scientist: a case study of the ambivalence of scientist. *Amer. Soc. Rev.*, 1974b, 39, 579-595.

Mitroff, I.I., & Mason, R.O. On evaluating the scientific contribution of the Apollo moon missions via information theory: a study of the scientist-scientist relationship. *Management Science: Applications*, 1974, 20, 1501-1513.

Mulkay, M.J. Norms and ideology in science. *Social Science Information*, 1976, *15*, 637-656.

Mulkay, M.J. *Science and the sociology of knowledge*. London: Allen and Unwin, 1979.

National Science Teachers Association, School science education for the 70s. *The Science Teacher*, 1971, *38*(8) 46-51.

Nay, M.A., & Crocker, R.K. Science teaching and the affective attributes of scientists. *Sci. Ed.*, 1970, *54*(1), 59-67.

Noll, V.H. The habit of scientific thinking. *Teachers College Record*, 1933a, 35, 1-9.

Noll, V. H. Teaching the habit of scientific thinking. *Teachers College Record*, 1933b, 35, 202-212.

Offner, M.F. Fact versus theory, *Sci. Educ.*, 1937, 21, 28-30.

Podolsky, B. What is Science? *The Physics Teacher*, 1965, 392), 71-73.

Popper, K.R. *The Logic of scientific discovery*. London: Hutchinson, 1968.

Roe, A. The psychology of the scientists. *Science*, 1961, *134*, 456-459.

Rothman, R.A. A. Dissenting view on the scientific ethos. *Br. J. Soc.*, 1972, *23*, 102-108.

Scheffler, I. *Science and subjectivity*. Indianapolis: Bobbs-Merrill, 1967.

Storer, N.W. *The Social System of Science*. New York: Holt, Rinehard & Wintson, 1966.

Sund, R.B., & Trowbridge, L.W. *Teaching science by inquiry in the secondary school*. Columbus, Ohio: Merill, 1973.

Swenson, L.S. The Michelson-Morley-Miller experiments before and after 1905. J. *Hist. Astron.*, 1970, *1*(1), 56-78.

Van Deventer, W.C. et al. Science for general education in the colleges. In Henry, 1960, pp. 97-111.

Vitrogan, D. Origins of the criteria of a generalized attitude towards science. *Sci. Educ.*, 1967, *51*, 175-186.

Vitrogan, D. Characteristics of a generalized attitude toward science. *School Science and mathematics*, 1969, *69*, 150-158.

Ward, C.H. The goals of high-school science. *Harvard Teachers Record*, 1933, *3*, 179-183.

West, S.S. The ideology of academic scientists. *I.R.E. Transactions on Engineering Management*, 1960, EM-7(2), 54-62.

Whipple, G.M. (Ed.). *The thirty-first yearbook of the National Society for the Study of Education. Part I: A program for teaching science*. Chicago: National Society for the Study of Education, 1932.

3

What is the Scientific Attitude ?

Robert L. Ebel

There are two reasons why a sound definition of the scientific attitude should be a part of the progressive science teacher's equipment. In the first place, modern educational thought regards the imparting of the scientific attitude as a fundamental obligation of science teaching. This conviction is clearly revealed in statements of objectives for science teaching and in the reports of individuals and committees concerned with science education. In the second place, on prerequisite to effective teaching of the scientific attitude is a concise definition of it. For we can neither teach effectively nor measure the results of our teaching without a definite notion of what is to be taught.

The problem of providing a sound definition of the scientific attitude has already received considerable attention at the hands of philosophers, scientists and educational research workers. Aristotle, Bacon, Huxèly, Darwin, Dewey and Russell are among the many who have attempted to analyze subjectively the attitudes of mind that contribute to scientific achievement. It is the rather popular scientific fashion to regard with contempt the products of such subjective thinking. Yet in this case that contempt is misplaced, for as Noll has pointed out, "Obviously the definition (of the scientific attitude) must depend on subjective analysis because attitudes cannot be observed directly but only through their manifestations in behavior." We may observe the behavior directly, but we can only speculate on the mental condition which produced it. Thus, whether we like it or not, our definition of the scientific attitude must be based on inferences from observed behaviour.

But it is not necessarily true that this subjectivity dooms any attempt at definition to incorrectness. The inferences need only be the starting point of our definition. If they are examined critically, subjected to such tests as can be devised, and combined judiciously, their inaccuracies may be exposed and eliminated. It is likewise not necessarily true that any one of the these philosophical definitions is complete and satisfactory in itself. And certainly they cannot each be complete and satisfactory, for in most cases they differ considerably from each other. Being the products of individual, opinions and hence being conditioned largely by the past experiences of the individual, they very widely in content and emphasis. It is altogether possible that no one of them is complete. It is probable that some of them contain incorrect elements. Therefore, while we must build our definition of the scientific attitude on the basis of these subjective judgments, it is apparent that we cannot use those judgements directly. Some technique for checking and compiling them must be employed.

Two recent efforts to define the scientific attitude illustrate on approach to this problem of checking and compiling the subjective judgements. The philosophy underlying the procedure in both cases apparently was something like this.

> The scientific attitude consists of a relatively small number of essential elements. But. the philosophical definitions of the attitude have suggested a great many elements. Some of these suggestions are probably wrong. Certainly not all of them are worthy of equal stress. Therefore, the job of the objective definer of the scientific attitude is to submit these multitudinous philosophic suggestions to competent judges, who will decide which, being most important, should be included in the final definition.

In both cases the same general procedure was followed. The investigator reviewed the philosophical definitions and prepared a tentative list of elements which in his opinion might be elements of the scientific attitude. He submitted this list to a number of selected judges, who evaluated the various suggestions and proposed additions, eliminations, and changes in the wording, emphasis or arrangement. The investigator then revised his list on the basis of the various judgements.

A comparison of the definitions produced by this method in the hands of the two investigators reveals a striking lack of agreement. While harmony between the two definitions would not be the only or even the best indicator of correctness, since consistent use of faulty technique could produce consistent though faulty results, it is true that the absence of substantial agreement does suggest a lack of soundness. Therefore, we are led to examine the procedure critically in the hope of discovering its faults.

We observe, in the first place, that the two investigators did not start with the same preliminary list of attitudes. There are very definite reasons why each investigator had to edit the philosophic suggestions concerning scientific attitudes before passing them on the judges. To have included all the suggestions that occur in the literature would have given an unreasonably long questionnaire. To have quoted the suggestions exactly would have given a confusing questionnaire, since the suggestions occur in varying contexts and under various forms of statement. But apparently it was felt that this preliminary editing was of minor importance when compared with the essential process of securing expert judgments, for no rigorous, definite editorial policy was followed. As a consequence, the preliminary lists were decidely different and, what is more important, that difference was now appreciably reduced by the modifications which the judges suggested. It is thus evident that the subjective preliminary editing which was almost entirely uncontrolled in the procedure used, was really of great significance in determining the nature of the final definition. The introduction of this important but uncontrolled factor in the process of definition constitutes the first defect in technique.

In the second place, we notice the questionnaire technique was employed to secure judgements which call for much reflective thinking. While there is apparently no specific objective evidence on the reliability of the questionnaire as a device for selecting elements of the scientific attitude, there is reason for general distrust of the questionnaire as a means of securing judgements. In an analysis of seventy-seven questionnaires studies involving expression of opinions Koos found that fifty-two were questionable on the ground that the persons responding lacked ability to give truly accurate answers.

Furthermore, there are several reasons why less careful answers are given to questions requiring the exercise of judgement. Reflective thinking is hard work and taken time. The experts to whom the questionnaires are addressed, being busy with their own endeavors not directly interested in the matter being probed by the questionnaire, will tend to answer in haste which precludes the exercise of the necessary reflective thinking. The questioner has no right to expect, and apparently does not often get thoroughly thought out answers to questions that require judgment. For these reasons, the use of the questionnaire in developing the definition constitutes second defect of technique.

The third weakness is that the procedure secures composite judgements in relation to a matter that really demands advance advanced reasoning. Is there any reason to believe that the composite judgment of a group of experts concerning scientific attitudes is necessarily closer to the truth than the judgement of any single member of the group ? In Galileo's day, would be the composite judgment of the professors concerning falling bodies have come closer to the truth than the judgment of one young professor in the group ? New ideas are not developed by composite judgements they arise from independent, individual investigation. The definition of the scientific attitude is a problem which requires the development of new ideal. Composite judgments may be interesting information, and useful in certain connections, but they should not be set up as ideals or standards. They represent the status quo, not the ultimate desideratum.

Finally, we notice that the definitions rest upon supposedly expert judgment, although there is, as yet, little expertness with reference to the scientific attitude. Expertness in any field is a specific thing, and no matter how well trained or successful a scientist or science teacher may be, he does not qualify as an expert in judging the elements of the scientific attitude unless he has specifically and thoroughly studied the nature of the scientific attitude for himself. There is little evidence in the literature that this specific and thorough study has been sufficiently wide-spread to create any significant body of experts. Furthermore, the exercise of expert judgement presupposes the existence of some commonly accepted standard or criterion as a basis for the judgment. But in this case no such standard was set up by the investigators, and it is unlikely that any substantial proportion of the judges took time to develop consistent standards of their own. It in this absence of expertness and of standards for judgment that we find the fourth defect in technique.

In our discussion of the procedures used by these investigators we have considered only the defects. The fact that there were four of them by no means proves that the definitions are valueless. On the contrary, they are among the best yet produced. But since the defects are serious and, with the possible exception of the first, are inherent in the technique itself, it is evident that any real improvement in the definition of the scientific attitude will require a different approach to the problem.

As a first step in attempting a new attack on the problem of defining the scientific attitude, we formulated the following basic philosophy.

> The scientific attitude is a unified state of mind. Philosophical speculations have suggested a great many characteristics of the attitude. Any sound definition of the scientific attitude must be based upon these suggestions. But some of the suggestions may be wrong. And in their present form they are confused, disjointed and overlapping. Therefore the job is the logical definer of the scientific attitude is (1) to search the literature for as many philosophical suggestions of the scientific attitude as possible, (2) to set up certain definite criteria for judging which are and which are not true characteristics of the scientific attitude, (3) to test these suggestions in terms of the criteria established and retain only those which pass the test, (4) to classify the characteristics accepted in terms of their common elements, (5) to determine the relations existing between the various classes and members of classes, and (6) to formulate finally an integrated statement of the scientific attitude which truly corresponds to the unified state of mind.

Perhaps the whole idea of this philosophy can be made clearer by the use of an analogy. Suppose that you discover a box containing a number of irregularly shaped pieces of cardboard. You recognize that some of the pieces are elements of a jig-saw puzzle, and immediately set to work to reconstruct the puzzle picture. Your first move, probably, would be to separate the jig-saw pieces from the scraps of cardboard, making use of the certain criteria, such as the shape of the piece and the coloration of the faces. Your next move would be to sort out the various pieces on the basis of their color, design, border, etc. Then you would begin to fit together the individual pieces in each group, and to relate each group to the others. The gradually appearing design would facilitate the work and would enable you to detect the absence of any piece and to determine its general character. In the end, a complete picture would be obtained, and you would be satisfied that you had solved the puzzle, not because someone approved your work, but because of its apparent internal consistency and completeness.

The process of constructing a sound, integrated definition of the scientific attitude is similar, but very much more difficult. The various suggestions constitute the pieces of our puzzle (although philosophic suggestions are considerably less definite than pieces of cardboard). The first task is to sort out the true elements of the scientific attitude from the false, using a definite standard or criterion. The next is to classify the various elements and to fit them together. In this process, certain characteristics of the attitude will gradually become apparent and will aid in the placement of other elements. Finally a complete, integrated definition will emerge.

Acting in accord with this philosophy, we set about to develop a satisfactory definition of the scientific attitude. The following requirements for the finished products were set up to define the problem and guide our work.

I. The statements must be complete, being based on the widest possible survey of suggestions concerning characteristics of the scientific attitude.

II. The statement must be accurate.

   A. All of its elements must fit a rigorous definition of the term "attitude".

   B. All of its elements must fit a rigorous definition of the term "Scientific" as it is used in relation to attitudes.

   C. All of its elements must be stated in concise, meaningful terms.

III. The statement must be integrated.

   A. All of its elements must be grouped on the basis of their common characteristics.

   B. All of its elements must be arranged to show the relationships existing between them.

On the basis of these requirements, the first problem was that of surveying the literature in order to gather as many suggestions concerning the scientific attitude as possible. The references which provided these suggestions are marked with an asterik in the list at the end of the article. A total of 432 separate suggestions were gathered. Not all of the suggestion were definitely labeled as elements of the scientific attitude in the context. If a statement seemed to relate in any way to the scientific attitude, it was "booked for investigation".

Furthermore, not all of the suggestions came from recognized authorities, for in the process of collecting, classifying and arranging these suggestions, certain other ideal occur to the investigator and were listen along with the rest. Such additions could be justified on the ground that it was not authority of source, but compliance with definitions of "attitude" and "scientific" which was to determine whether or not the suggestion should be accepted. To aid in a later process of classification and arrangement, each suggestion was given a label which identified its source and expressed in central idea, and was then typed on a separate slip of paper.

With the suggestions collected, the next problem was to train a rigorous definitions of "attitude." The definitions given in six standard dictionaries, one encyclopaedia, and in one other reference were collected and carefully studied. It soon became apparent that the term "attitude" has at least three different meanings, which are not always distinctly separated in the definition given. There are physical attitudes, expressive attitudes, and mental attitudes. It is clearly this third meaning which relate to the scientific attitude.

But even with this distinction established there remain an apparently irreconcilable difference between the various definition which relate specifically to the mental attitude. If a mental attitude is a "position of disposition with regard to a person of thing," then it is obviously a mental condition. If, on the other hand, a mental attitude as a "habitual mode of regarding anywhere," then it is clearly a type of mental activity. The original use of the term in applying to physical position (or condition) lends weight to the first interpretation. Furthermore, the most authoritative definition of a mental attitude on the list, since it was proposed by the American Psychological Association, supports the mental condition concept by defining an attitude as "a catalized set or disposition." If an attitude as a mental condition, it is not difficult to account for the confusion which has caused to be regarded as a mental activity. For a mental condition cannot be observed directly, but only through its influence on behaviour. Thus the behaviour has been confused with the mental condition which produced it. For all these reasons we are led to favor the mental condition concept.

On the basis of these distinctions and interpretations, the suggestions of the various definitions were combined into this final statement which was adopted as our guide:

An attitude is a stabilized mental set which expresses itself in a tendency to react to any member of a class of stimuli in the same general way.

The term "stabilized" expresses the relative permanence of attitudes. Attitudes are, in a sense, habitual, but it is hardly accurate to define them as "habits of thinking". For habits of thinking might include such things as "Two plus two equals four," and "Where there is smoke, there is fire," which are certainly not attitudes.

The subordinate clause in the definition is simply an explanation of the meaning of "mental set," in terms of the effect it produces. Such an explanation is necessary, because in using the definition to pick out mental sets, the sole basis of judgment is the effects which the sets produce. The latter part of the definition also expresses the generalization nature of attitude in the term, "class of stimuli." If, for example, a person has an attitude of fear toward snakes, that attitude functions whether the stimulus is a large snake or a small one, a live snake or a dead one, a poisonous snake or a harmless one. In fact, the stimulus need not even be a snake at all, for an artificial imitation, or even a suspiciously sinuous stick of wood, will evoke the same type of response.

The next problem was a formulate a rigorous definition of the term "scientific" as it is used in relation to attitudes. Scientific attitudes might be interpreted either as the attitudes which have existed in the minds of outstanding men of science, or as the attitudes which tend to foster scientific achievement. The two interpretations overlap, but are not identical. Outstanding scientists have frequently displayed such attitudes as those of appreciation and reverence, which are common characteristics of well-developed personalities, but which do not tend to foster scientific achievement. It is obvious that the second interpretation will give us a list of attitudes that is more thoroughly scientific and more useful. On the basis of these considerations, the following definition of a scientific attitude was formulated.

A scientific attitude is an attitude which will tend to foster scientific achievement.

The term "scientific achievement" is here used in a broad sense and includes: (1) any addition to the world's store of organized truth, (2) any addition to the individual's store of organized truth, or (3) any use of organized truth as a basis of determined action.

A word should be said here about the distinction between the scientific attitude and the scientific method. The two terms have apparently been much confused in the past, although they really involve quite separate concepts. An attitude is a mental condition, while a method is an organized series of acts. The scientific attitude is indeed closely related to the scientific method, for the attitude gives rise to the method, and the method gives evidence of the attitude. It is this close relationship which has led to the confusion.

The accepted elements of scientific attitude were classified in terms of their common characteristics and arranged in accordance with the relations existing between them. In stating each element an effort was made to use clear, concise terminology. Because of the fact that there are not separate names available for each of the elements of the scientific attitude, it was necessary to refer to them in terms of their influence on behaviour. Hence the word "readiness" is used here to indicate a mental set which inclines the individual to certain types of behaviour. Thus the statement, "Readiness to be open-minded", might be translated as, "A mental set which inclines the individual to be open-minded". "Open-mindedness" describes the behaviour; "Readiness to be open-minded" describes the attitude. The statement follows.

The scientific attitude includes :

I. Readiness to be confident means that human intelligence can understand the phenomena of nature and through that understanding can control the forces of life. This element of the scientific attitude—

A. Consists of

1) Readiness to be confident that natural phenomena are understandable, which involves --

   a. Readiness to be confident that the Universe is a self-sustaining unit, which includes --

i) Readiness to believe that there is no supernatural interference in the universe.

ii) Readiness to believe that even man is a natural phenomenon.

iii) Readiness to believe that even man is a natural phenomenon.

b. Readiness to be confident that there are orderly interrelations between all natural phenomena, which includes --

i) Readiness to be confident that there are orderly.

a) Readiness to believe that there are reasons for all things, including -

i. Readiness to believe that the same cause always produces the same effect.

ii. Readiness to believe that consequences can be deduced from or explained in terms of their antecedents.

ii) Readiness to believe that recurring sequences of cause and effect indicate permanent general relation which may be formulated on natural laws, involving-

a) Readiness to believe that the truth itself does not change, although our ideas of it may change.

b) Readiness to believe that these laws may be used to accurately predict future occurrences of natural phenomena.

c) Readiness to be confident that these various relations indicate a systematic unity in natural phenomena.

2. Read to be confident that human intelligence is capable of understanding natural phenomena, which includes -

a) Readiness to have confidence in sensory data, which includes.

i) Readiness to believe that the world of common sense experience cannot be transcended.

ii) Readiness to believe that the objective evidence obtained by observation of the external world is the only dependable source of truth.

b) Readiness to be confident that this understanding will

enable man to gain increasing ability to adjust himself to the forces that affect life, and to gain increasing control over those forces.

B. Gives rise to --

1. Readiness to discredit and abandon all beliefs which depend upon supernatural interference for the suspension of natural law, which involves
   a) Readiness to discredit and abandon superstitious beliefs.
   b) Readiness to discredit and abandon beliefs in miraculous occurrences.
   c) Readiness to discredit and abandon belief with the hope of getting something for nothing.
   d) Readiness to discredit and abandon belief in the innate authority of hunches, first impressions, feelings, etc.
2. Readiness to accept tested human knowledge.
3. Readiness to deny the possibility of danger in flooding the world with too much acknowledge.
4. Readiness to seek personally and to understand the phenomena of nature, which involves--
   a) Readiness to seek a unified, orderly comprehension of natural phenomena.
   b) Readiness to rely ultimately upon individual thinking and independent judgement.
5. Readiness to attack problems with reason, which involves-
   a) Readiness to think freely, defying all dogmas, precedents, authoritative dicta and traditions which are not based on reason.
   b) Readiness to seek truth through observation, experiment and thinking.
   c) Readiness to look for true and cause and effect relations.
   d) Readiness to use imagination constructively to suggest a variety of hypothetical solution to a problem.
   e) Readiness to test the hypotheses in the hope of developing

dependable laws which will solve not only the immediate problem but also other problems of like nature.

6. Readiness to deny that the status quo or the trend of events is inevitable.
7. Readiness to seek to control human action with persuasion, not with force.

II. Readiness to seek true understanding of the phenomena of nature. This element of the scientific attitude -

A. Arises from-

1. Readiness to love knowledge for its own sake.
2. Readiness to desire improvement in the status quo.
   a) Readiness to seek increasing control over the force, not for the private exploitation of other humans, but for the progressive advancement of social welfare.
3. Readiness to seek personal achievement.

B. Gives rise—

1. In relation to the phenomena of nature in general, to —
   a) Readiness to have broad and versatile interests.
   b) Readiness to be sensitively curious.
   c) Readiness to attempt the discovery of new knowledge.
2. In relation to specific problems, to —
   a) Readiness to make an active attack on the problem.
   b) Readiness to be industrious, persistent and relentless in the search for truth.
   c) Readiness to overcome obstacles for the discovery of truth, which involves—
      i) Readiness to face personal danger in the pursuit of truth with a courageous, adventurous spirit.
      ii) Readiness to defy intolerance, to cast-off restrictions and to think freely in the pursuit of truth.
      iii) Readiness to sacrifice personal interests, profit, reputation, or even life itself, with an unselfish, heroic devotion to truth.
      iv) Readiness to disregard any unpleasantness of condi-

tions, processes or conclusions in search for truth.

v) Readiness to ignore the apparent uselessness of the truth being sought, and to ignore the lack of assured material reward for its discovery.

vi) Readiness to attempt to improve the methods of obtaining truth.

vii) Readiness to record any openly published discoveries and developments so that they may make an unselfish contribution to scientific advancement and social progress.

3. In relation to scientific achievements, etc., to —

   a) Readiness to consider the disclosure to truth to be the greatest of all achievements.

   b) Readiness to respect and admire those who have contributed to the disclosure of truth.

III. Readiness to seek correctness in work and thinking so that truth may be discovered. This element of the scientific attitude includes—

A. Readiness to seek specific training for any observing, experimenting, thinking, or to other operation necessary to the discovery of truth.

B. Readiness to seek a factual basis for all conclusions, and to avoid assertion. This involves —

1. Readiness to distinguish between facts, which correspond to objectives of reality and fancies.

2. Readiness to investigate and gather evidence, which includes—

   a) Readiness to observe and to experiment.

   b) Readiness to read reports to similar investigations to acquire the data that others have gathered.

   c) Readiness to recall pertinent information out of all previous experience.

3. Readiness to examine and compare the data which includes—

a) Readiness to pick out essential element in the data.
b) Readiness to discover similarities, dissimilarities and exceptions in the data.
c) Readiness to determine what conclusion may be derived from the data.

C. Readiness to be carefully and painstakingly accurate in all work and thinking, which involves —

1) Readiness to define the problem.

2) Readiness to observe accurately, which includes —

a. Readiness to control the conditions under which observation takes place, including —

i. Readiness to isolate the facts to be observed, removing irrelevant objects and distractions.

ii. Readiness to avoid complexity and brevity in the events observed.

b. Readiness to observe minutely.

c. Readiness to use mechanical aids to extend the range of observation.

d. Readiness to use measuring instruments to improve the precision of observation.

3) Readiness to experiment accurately, which includes—

a. Readiness to experiment, to plan, and to execute the experiment rigorously and carefully.

b. Readiness to control conditions.

c. Readiness to use control experiments.

d. Readiness to permit only one variable.

4) Readiness to think accurately, which includes —

a. Readiness to think logically, involving—

i. Readiness to think sound, pertinent and adequate data in making inductions.

ii. Readiness to follow rules of logic in making deductions.

b. Readiness to think objectively, which involves—

i. Readiness to avoid rationalization.

ii. Readiness to avoid wishful thinking.

c. Readiness to think deeply, which involves—

i. Readiness to penetrate external appearances.

ii. Readiness to avoid basing conclusions on coincidence.

iii. Readiness to avoid basing conclusions on superficial analogy.

d. Readiness to think coldly, which involves —

i. Readiness to be impersonal and disinterested in thinking.

ii. Readiness to be unemotional, dispassionate and thoroughly self controlled in thinking.

iii. Readiness to control imagination.

iv. Readiness to avoid being swayed by more novelty or sensationalism of an idea.

5) Readiness to calculate accurately.

6) Readiness to use words accurately, which involves-

a. Readiness to read carefully.

b. Readiness to define all terms.

c. Readiness to define all terms.

7) Readiness to be accurate in statements, which involves-

a. Readiness to avoid assertion.

b. Readiness to avoid exaggeration.

c. Readiness to use quantitative expression.

8) Readiness to record procedure and results to the exactness of memory.

D. Readiness to be orderly in all work and thinking, which involves—

1) Readiness to be systematic in tackling a problem, in observation, and in examination of data.

2) Readiness to organize data and arrange it congently.

E. Readiness to seek completeness in all work and thinking, which involves —

1) Readiness to seek all the facts, which includes—

   a. Readiness to observe and experiment extensively and under and variety of condition.

   b. Readiness to read widely.

2) Readiness to consider all possible hypotheses.

3) Readiness to keep alert for and listen to any suggestion for further observation, experimentation, reading or thinking.

4) Readiness to persist until an adequate explanation has been discovered, or the whole truth made clear.

F. Readiness to avoid needless complexity in conclusions, which involves.

1) Readiness to favour the simplest explanation of a phenomenon.

2) Readiness to favour the most fruitful hypothesis which, having widest application relates the greatest number of facts of each other.

G. Readiness to be open-minded, which involves —

1) Readiness to recognize valid new conceptions, regardless of their conflict with tradition.

2). Readiness to prevent any doctrine from becoming a master to thinking.

3) Readiness to record and openly publish all discoveries and developments so that they may checked by others.

J. Readiness to suspend judgement until all facts are in and have been considered, which involves —

1) Readiness to refuse snap judgements and hasty generalizations.

2) Readiness to deliberately reflect on the facts.

3) Readiness to be active but patient in attempting to arrive at judgements, either expecting nor demanding immediate success in attempts to discover truth.

4) Readiness to tolerate and respect another's point of view, pending suitable investigation.

5) Readiness to recognize the tentativeness of all conclusions, including scientific conclusions.

6) Readiness to hold opinions tentatively.

K. Readiness to be critical of all work and thinking, which involves—

1) Readiness to examine all work and thinking to see if it has the characteristics of scientific correctness.

2) Readiness to extend the attitude of criticalness so that it includes self-criticism.

L. Readiness to be rationally and impartially sceptical of the products of human work and thinking, which involves —

— Readiness to be cautions and conservative in accepting conclusions.

M. Readiness to be intellectually honest, which involves —

1) Readiness to eliminate prejudices, bigotry and bias.

2) Readiness to be convinced by the evidence which involves —

   a. Readiness to admit error when it becomes apparent.

   b. Readiness to abandon pet hypotheses.

   c. Readiness to change opinion of the basis of new evidence.

3) Readiness to recognize the limits of accuracy, probability of correctness, and significant of all conclusions, including scientific conclusions.

4) Readiness to be humble, which involves —

5) Readiness to check and verify all work and thinking, which includes —

   a. Readiness to verify observations, experiments and thinking by repetition and comparison —

   b. Readiness to very deduction by checking them in terms of the rules of logic.

   c. Readiness to test hypotheses by applying them to the facts.

   d. Readiness to test theories by making predictions on the basis of them and comparing the predicted result with the observation result.

   e. Readiness to avoid gullibility and over credulity.

4
Scientific Attitudes

Richard H. Lampkin, Jr.

In most of the contemporary formulations of the objectives of instruction in secondary school science the attainment of both the scientific method of thinking and the scientific attitudes has been stressed. But, despite the coordinate importance assigned to the two, the scientific attitudes have been less well defined than the steps in the scientific method. Nevertheless, because the importance of the scientific attitudes has been generally recognized, a number of studies have been based on the existing incomplete formulations of them. Such studies have included attempts to develop, and to evaluate, methods and materials for teaching the scientific attitudes, to develop tests of the possession of them, and to determine any effects on the behavior of persons learning them. Such studies are undoubtedly desirable, but they seem to have been premature. They might have been more valuable had they been based on a more nearly complete definition of the scientific attitudes.

THE PROBLEM

The primary purpose of the study was to discover, and to state in definite and unambiguous propositions, as many as possible of the working assumptions and rules of practice which have been termed scientific attitudes. A secondary purpose was to discover the extent to which these scientific attitudes had been recognized in the professional literature of general science teaching.

A new definition of a scientific attitude is implicit in the above statement of the problem. It seems quite inconsistent with many of the proposition derived in the study of carry over into science the psychological definition of an attitude as a "feeling tone", or as "stabilised set or disposition" or as a "habit of thinking". Neither are scientific attitudes to be regarded as scientifically desirable forms of "emotionalized attitudes".

A scientific attitude may be either of two different things. It may be an assumption made as a basis for work, such as the assumption that anyone is liable to make mistake (Cf., Attitude 75, below). This proposition is compatible with past experience, but for the future it is only probable, an assumption. Or, a scientific attitude may be rule of practice, such as the rule that every effort must be made to prevent, and to discover and correct, possible mistakes (Cf. Attitude 76, below).

Scientific attitude are assumption and rules of practice which have been formulated consciously and deliberately, or at least have been subjected to searching criticism since their formulation. They are neither sacrosanct nor necessary. They are convenient and fruitful only. And blind habituation to them, or any emotion attached to them, is undesirable, for it might prevent the discarding of any which conceivably might prove no longer useful. Scientific attitudes are simply elements in the philosophy of science.

METHOD

The first part of the study considered of a critical analysis of twelve published works of the philosophers of the last half-century. The scientific attitudes so found were studied, and reported, by classifying them as to their applications to the three steps of the scientific method.

In the second part of the investigation 17 related studies were analyzed to determine which of the scientific attitude derive above they had recognised.

The philosophical selected for study were :

H. Poincare	:	The Foundation of Science
E. Mach	:	The Science of Mechanics

B. Bavink	:	The Natural Sciences
G. Santayana	:	Reason in Science
J. Deway	:	The Quest for Certainty
Good, Bars and Scates	:	The Methodology of Educational Research
Cohen and Nagel	:	An Introduction to logic and Scientific Method
K. Pearson	:	The Grammer of Science
J. Huxley	:	II'hat Dare I Think
B. Russell	:	The Scientific Outlook
F. Westaway	:	Endless Quest
A. Woli	:	The Essentials of the Scientific Method

The authors of the related studies in the field of general science teaching were: Twiss, Meister, Eikenberry, Curtis, Craig, Downing, Rice, Davis, Hunter and Knapp, Beanchamp, Skewes, Noll, Watkins, Bickel Caldwell and Crowell.

FINDINGS

The outcome of the first part of the study was the definition of ninety-two scientific attitudes. They are presented here in eighteen groups subordinate to the three steps of the scientific method.

Applicability of the Scientific Method

1. The scientific method is applicable to any materials that can be reasoned on.
2. The scientific method must not be barred from any materials that can be reasoned on.
3. The scientist is not interested in any materials that can not be reasoned on.

The first step in the scientific method is the experiencing of sense-impressions.

Sensitivity of Perception

4. The scientist must be responsive to sense stimuli.

5. The Scientist must keep his sensitivity to stimuli as high as possible.

Accuracy of Perception

6. The scientist must be accurate, i.e., his response must be in some consistent way proportional to the stimulus.
7. The scientist must guard against anything that might make his response to sense stimuli inaccurate.
8. The possibility of agreement between different observes depends upon the similarity of their perceptive faculties.

Objectivity of Perception

9. All measurement is relative to an accepted standard.
10. The scientific must compare any thing being measured as directly as possible with the standard, not indirectly through his memory of it.
11. The use of instruments permits a great increase in both the sensitivity and accuracy of observation.

Limitation of Perception

12. It is highly probable that there are phenomena no observer has sensed yet.
13. No one is able to make even an outline of all possible future developments in science.

The second step in the scientific method is the classification of sense impressions on some basis, that is, the formulation of scientific laws and hypotheses, with sense-impressions, as well as with data of greater complexity, this process is called generalization.

14. Without sense-impressions there would be nothing to be known.
15. The materials of science have no immediate sense-impressions, but rather interferences based on such sense impressions.

Scientific Classification

16. The scientist is interested in a particular event only in so far as it is an example of a class of phenomena.
17. Scientific classification, by grouping phenomena, facilitates the comprehension of them.

18. Scientific classification makes possible the concise recording of experience.
19. Scientific classification makes the education of the uniformed easier.
20. Statistics is very helpful in comprehending numerical data.

Sampling

21. Whatever is true of a fair sample will be found to be true of the class from which the sample is taken.
22. Only if a sample is truly representative of a class can be the nature of the class be inferred from that of the sample.
23. No characteristics found hitherto associated only with a nervous system may be inferred where no such nervous system has been found.
24. No one can reason across a breach in the routine of experience.

Nature of Scientific Law

25. From his experience a scientist may predict that if certain complexes of conditions exist, certain other complexes of conditions will, or do, exist.
26. Only if past experience is a fair sample of future experience will scientific laws based on past experience be valid in the future.
27. A scientific law expresses a relation between only certain elements abstracted from experience.
28. A scientific law can say just what the implication is between a "cause" and its "effect".
29. A scientific law does not imply any compulsion exerted by a "cause" and its "effect".
30. A scientific law does not imply any teleological connection between a "cause and its "effect".
31. A scientific law does not refer to concatenation of natural phenomena considered as outside and independent of the scientist.
32. A scientific law does not "rule nature".
33. A scientific law is not a transcription of some prodigious code prescribed by some higher power.

34. A scientific law exists only when it is formulated by the scientist.

Nature of Hypotheses

35. A hypothesis is assumption of something beyond sense-impression for the sake of correlating sense-impressions.
36. The formation of any hypothesis is logically invalid.
37. Sensations are not signs of things; but on the contrary a thing is a thought symbol for a compound sense impression of relative fixedness.
38. Any given sense-impression can be explained by more than one hypothesis.
39. There is no sharp distinction between fact and hypothesis.
40. It is unnecessary, as well as unjustified, to hope that a hypothesis represents any thing beyond sense-impression.
41. The mutual incompatibility of some hypothetical conception in science results from their not being sufficiently abstract.
42. It must never be forgotten that a hypothesis correlates only abstract elements of experience.

Formation of Laws and Hypotheses

43. There are no known rules that can be followed mechanically to arrive at useful hypotheses.
44. Not all possible hypotheses can be enumerated.
45. The knowledge of a field of subject matter is conducive to the formation of useful hypotheses in that field.
46. Inspiration, or intuition, is essential to the formation of hypotheses.
47. Valuable laws and hypotheses spring more than from a spring of curiosity, or of inquiry for its own sake, than from hope of financial gain, or of increased prestige.
48. Ultimately the direct influence of pure science or practical life is enormous.
49. Critical dissatisfaction with thing-as-they-are is a more productive scientifically than admiration.

The third step in the scientific method is checking the stated relation in the scientific law, or the implied relations in the hypotheses, against sense-impression to make sure that such relations exit. Stated differently the third step is the elimination of incompatible laws and hypotheses.

No possible Proof

50. It is a fallacy to "affirm the consequent".
51. The "argument from design" is not valid.
52. No law, or hypothesis, can be proved; it can only be shown to be compatible with sense-impressions.
53. Of the hypothesis compatible with the facts, the best hypothesis is that which is the systematically simplest one.

Compatibility with Accepted Concepts.

54. The statement of law, or hypothesis, must be clear and unambiguous.
55. All assumptions made must be clearly stated as such.
56. A hypothetical proposition must be differentiating; it must not explain no-matter-what may happen.
57. A hypothesis must account explicitly for all relevant concepts which are accepted at the time of its formulation, either by corroborating or by discrediting them.
58. Results taken from other workers must be criticized as searchingly as are one's own.
59. Whenever a law, or a hypothesis with its logical implications, is not compatible with sense-impressions, it must be revised until it is compatible.
60. When his hypotheses and laws are shown to be inadequate the scientist revises them to form closer and closer approximation, rather than discards them to start completely anew.
61. A hypothesis need not be true to produce valuable results.

Test by Experiment

62. The one most important check on itself the scientific method has is the over-all check against sense-impressions.
63. If the data of record relevant to a proposed law, or hypothesis, are insufficient for judging it, additional data must be gathered

by an experiment designed to test it.

64. If two complexes of conditions, C and E, do not vary concomitantly, C and E are equally related.
65. It is fallacy to assert that if two complexes of conditions, C and E, do vary concomitantly, the C and E are casually related.
66. In order to establish a concomitant variation between C and E, the ability to measure both C and E is necessary, and statistical methods of handling the data as desirable.
67. If C ever occurs without E, than C and E are not casually related.
68. If C occurs with E, then C and E casually related.
69. If an experiment all relevant independent variables, but one should be held constant.
70. It is impossible for the scientist to hold constant all independent variables but one.
71. The relevance of any variable can be determined only by an experiment.
72. The effect of two "causes" acting simultaneously is calculable from the effects they have separately.
73. Successful prediction of a phenomena, if fields are different from that in which it was derived, is a strong evidence for a hypothesis.
74. It is a fallacy to argue that a proposition is true simply because it has not been or cannot be, proved false.

Guarding against error

75. Every one is liable to make mistakes.
76. The wise scientist will use every conceivable means to prevent and to discover and correct, possible errors in his work.
77. The effects of unavoidable errors of observation may be minimized by applying to them the theories of probability.
78. Failure to correct errors often gives rise to superstitious beliefs.
79. The scientist attempts to form judgements which will be independent of the idiosyncrasies of his individual mind.
80. The scientist deliberately renounces all emotion and desire, except that of accomplishing his two fold purpose.

81. The scientist should neve make pontifical announcement for indulging in melodrama.
82. If the scientist is ignorant of certain points, he must acknowledge his ignorance of those points.
83. The scientist's acknowledge of his present ignorance is not a resignation to defeat; it is rather that which leaves open the way for future investigation.

Persistent Search for Adequate Conceptions

84. A hypothesis must not be accepted so long as other hypotheses have probabilities of the same order.
85. The search for adequate laws and hypotheses must persist not only in one man but in generation of men.
86. There are no abstract principles by which the nature of things can be predicted before investigating them.
87. No concept, no matter how it was arrived at, should be adopted until it has been shown to be compatible with sense-impression.

Mind Open to Necessary Change

88. There is not idea so firmly established that it cannot some time need to be changed.
89. More intensity of belief in a proposition is no guarantee of its truth.
90. Neither custom nor tradition can guarantee the validity of any conception.
91. No authority can guarantee the validity of any conception.
92. No scientist can slough the responsibility for making his own decision.

The second part of the study revealed that there had been fairly general recognition and acceptance of the scientific attitudes grouped under the following topics.

1. Applicability of the scientific method.
2. Limitations of perception.
3. Materials of science are drawn from sensitive, accurate, and

objective perception.

4. Hypotheses and laws must be compatible with accepted concepts.

5. Test by experiment.

6. Guarding against error.

7. Persistent search for adequate conceptions.

Also, it was shown that the attitude under the following heads had been neglected almost completely, namely,

1. Laws and hypotheses are based on sampling of experience.

2. No conception can be demonstrated to be necessarily true.

Similarly, there were other attitude which had been recognised after a fashion, but evidently had been understood in senses quite different from those developed in this study. Such attitudes as included in the groups under the following categories :

1. Nature of scientific law

2. Nature of hypothesis

3. Formulation of laws and hypotheses

4. Necessity of renouncing emotion in scientific work.

5

Scientific Attitude

M. S. Yadav

One of the major aims of teaching science is the development of scientific attitude. Following are some of the various aspects included in scientific attitude.

(i) Making pupils open-minded.

(ii) Helping pupils make critical observations.

(iii) Developing intellectual honesty among pupils.

(iv) Developing curiosity among pupils.

(v) Developing unbiased and impartial thinking.

(vi) Developing reflective thinking.

NSSE (National Society of the Study of Education) has defined scientific attitudes as "open mindedness, a desire for accurate knowledge, confidence in procedure for seeking knowledge and the expectation that the solution of the problem will come through the use of verified knowledge."

The views regarding scientific attitude expressed at the workshop conducted in the National Council of Education Research and Training (NCERT) at Chandigarh in 1971 can be summarised as follows. A pupil who has developed scientific attitude

(i) is clear and precise in his activities and makes clear and precise statement.

ii) always bases his judgement on verified facts and not on opinions.

iii) prefers to suspend his judgement if sufficient data is not available.

iv) is objective is approach and behaviour.

v) is free from superstitious.

vi) is honest and truthful in recording and collecting scientific data.

vii) offer finishing his work taken care to arrange the apparatus, equipments, etc., at their proper places.

viii) shows a favourable reaction towards efforts of using science for human welfare.

6
The Scientific Attitude

Victor H. Noll

The scientific attitude includes the following habits of thinking.

1. Habit of accuracy in all operations, including accuracy in calculation, observation and report.
2. Habit of intellectual honesty.
3. Habit of open-mindedness.
4. Habit of suspended judgement.
5. Habit of looking for true cause and effect relationship.
6. Habit of criticalness, including that of self-criticism,

The titles as listed are probably sufficient to describe the habit in detail.

They may, however, be brief defined for further clarity in terms of their opposites:

1. Accuracy in calculation, observation, and report in the opposite of habits of careless inaccurate work.
2. Intellectual honesty is the opposite of such habits as exaggeration and rationalization.
3. Open-mindedness is the opposite of bigotry, prejudice and intolerance.
4. Suspended judgement is the opposite of the habit of making snap judgements or of jumping to conclusions.

5. Looking for true cause and effect relationship is the opposite of habits of superstitious thinking, of expecting rewards to come without commensurate effort.

6. The habit of criticalness including that of self-criticism is the opposite of habits of accepting explanation of phenomena without question, or without attempt at evaluation; it is the opposite of the habit of condoning and accepting such things as racketeering political corruption, and the like, as inevitable.

Fourteen specific objectives are listed here without attempt to rank them. The fourteen specific objectives are --

1. Command of factual information.
2. Familiarity with laws, principles and theories.
3. Ability to distinguish between fact and theory.
4. Concept of cause and effect relationship.
5. Ability to make observation.
6. Habit of basing judgement on fact.
7. Ability to formulate workable hypotheses.
8. Willingness to change opinion on the basis of new evidence.
9. Freedom from superstitious.
10. Appreciation of the contribution of science to our civilization,
11. Appreciation of natural beauty.
12. Appreciation of man's place in the universe.
13. Appreciation of the possible future developments of science.
14. Possession of interest in science.

7
Scientific Attitudes

R. C. Sharma

Scientific attitudes are the most important outcomes of science teaching. Though some people view the scientific attitudes as the by-products of teaching science, yet a majority of the people consider them as equally important as the knowledge aim. Science attitude has a number of characteristic feature which distinguish it from other attitudes.

A man with scientific attitude

(i) Is critical in observation and thought.

(ii) Is open- minded.

(iii) Respects other's points of view and is ready to change his decisions on presentation of new and convincing evidence.

(iv) Is curious to know more about the things around him. Wants to know 'Why', 'Whats' and 'Hows' of the things be observes.

(v) Is objectives in his approach to problems.

(vi) Does not believe in superstitions and false beliefs.

(vii) Suspends judgements until suitable support is obtained.

(viii) Believes in cause and effect relationship.

(ix) Is truthful in his observations and draws conclusions based on accurate facts.

(x) Is unbiased and impartial in his judgements.

(xi) Adopts a planned procedure in solving a problem.

(xii) Believes that truth never changes, but his ideas of what is true may change as he gains better understanding of that truth.

(xiii) Accepts no conclusions as final or ultimate.

(xiv) Seeks to adopt various techniques and procedures to solve the problem.

(xv) Selects the most recent, authoritative and accurate evidence related to the problem.

(xvi) Seeks the facts and avoid exaggeration.

8
Scientific Attitude

Otis W. Caldwell
&
Francis D. Curtis

1. A curiosity to know about one's environment.
2. The belief that nothing can happen without a cause and that occurences that seem strange and mysterious can always be explained by natural causes.
3. An unwillingness to accept as facts any statements that are not supported by convincing proof.
4. The determination not to believe in superstitions of any sort.
5. The belief that truth itself never changes, but that our ideas of what is true change as we gain more and more knowledge.
6. An intention not to experiment or to work blindly and carelessly, but to begin only after careful planning.
7. The determination to be careful and accurate in all one's observations.
8. A willingness to consider all the evidence and to try to decide whether it really relates to the matter which is being considered, whether it is sound and sensible, and whether it is complete enough to allow a conclusion to be made.
9. A determination not to base final conclusions on one or a few observations, but to work as long as may be necessary in order to secure an answer to a problem.

10. The desire to do one's own observing an dexperimenting, but a willingness to use the results of other scientist's work.
11. The willingness to change an opinion or a conclusion if later evidence shows that it is wrong.
12. The intention to respect another's point of view.
13. The determination to prevent one's own likes and dislikes from influencing one's judgment.

9

Emphasise Scientific Attitudes in Teaching and Learning of Science

Nathan S. Washton

Scientific attitudes must be included in teaching and learning of science. The scientific attitudes are

1. Open-mindedness-willingness to consider new facts.
2. Intellectual honesty—scientific integrity, unwillingness to compromise with truth as known.
3. Senstivity to possible uses and applications of science in personal relationships and disposition to use scientific knowledge and abilities in such relations attitudes.

10

Components of the Scientific Attitude

Paul B. Diederich

For some purposes of identification, some of the components of scientific attitude or attitudes are as follows:

1. *Scepticism. Not taking things for granted. Asking the prior question.* Must of our thinking takes the form: "Science, as everyone knows, X is true, Y must follow." The average man focuses his attention on the latter part of the enthymeme and argues that Y does not necessarily follow. The scientist characteristically asks: "But how do we know, after all, that X is true?
2. *Faith in the possibility of solving problems.* Most people believe that there are problems that human intelligence will never solve; there will always be war, poverty, ignorance sickness, and all sorts of misery. A scientist would tend to look at these things as problems that we can solve—a little bit at a time. The important thing is to see them as problems.
3. *Desire for experimental verification.* The average man, having once decided that something is true, has no desire to carry the matter further, except possibly to argue about it. The Scientists want to submit the idea to experimental proof. There is a further step in this process that possibly we should focus upon. Practically no great idea can be tested experimentally as it

stands; it is too complex, and the conditions to which it has reference cannot be manipulated. One first has to make a series of inferences: "if this idea is true, then A, B, C, D..... N must follow." Perhaps, of this series, only L and M can be tested experimentally, and even if L occurred, some other hypothesis might explain why it occurred. Very well, then, we test M.

4. *Precision.* A scientist is impatient with vague, wooly, emotional statements of what given observations or experiences mean. This is a trait that seems to be relatively easy to test objectively.

5. *A linking for new things.* Most people like to be old-fashioned and unscientific. Suggest any idea that is new to them, and they will think of more unsound or cannot be done than you could answer in a day. Some scientist has to get hold of the idea and demonstrate that it is sound, workable, and helpful before the unscientific will even begin to consider it seriously. By that time the scientist has a patent on it.

6. *Willingness to change opinions.* People hate to give up an idea—especially one they thought clever, or sacred, or a part of the general structure of ideas on which their security depends. A scientist feels these pangs also, but is more willing than most to alter an opinion, once he sees reliable evidence to the contrary, because he knows that, every time he does so, he has learned something. Retaining the old opinions intact is satisfying to the ego, but a sure indication that has learned nothing.

7. *Humility.* Most people are extremely arrogant in their opinions. In any tavern—and even on a good many lecture platforms—one will hear opinions expressed with supreme confidence that could not be tested or proved. A scientist realizes how little is known with any certainty; he commonly looks for little truths that the unscientific would consider not worth the trouble.

8. *Loyalty to truth.* A scientist is sometimes subjected to humiliations as his findings shift and invalidate some conclusion to which he has previously committed himself, but his loyalty to truth is such that he would rather cut off his right arm than suppress the new data. The general picture is of lofty and even noble devotion to the facts, however they may affect one personally.

9. *An objective attitude.* A scientist has a high regard for facts and tries to behave in accordance with them, while an unscientific person tends to see only the facts he wishes to see and to react emotionally against others. Leo Szilard (One of the top men working on the Bomb) said that just before the last war, intelligent Englishmen were convinced that war was about to break out, but they behaved as though this possibility had nothing to do with their everyday conduct. One such person was showing Szilard his new apartment, very luxuriously furnished. " What do you think of it?" he asked. "It's combustible!" said Szilard.

10. *Aversion to superstition.* Although we tend to think to this battle as so nearly won that it is no longer worth serious attention, astrologers and faith healers still flourish, and most of the children in an average high school are riddled with superstitions.

11. *Linking for scientific explanations.* People tend to like mythical or romantic explanations of phenomena; the coldly factual explanation strikes them as taking all the glamor away. We may have to provide the substitute satisfaction, even though it is rather snobbish, of knowing things that not generally known or believed.

12. *Desire for completeness of knowledge.* A scientist is often impelled to round out his knowledge of a subject; to fit every piece into its place like a jig-saw puzzle. The missing piece torture him. Yet he is more willing to live with this incompleteness than to fill in the gaps with off-hand explanations.

13. *Suspended judgement.* As scientist tries hard not to form an opinion on a given issue until he has investigated it, because it is so hard to give up opinions already formed, and they tend to make us find the facts that support the opinions. This is closely related to a desire to investigate before acting—to get all the relevant facts if immediate action is not necessary. There must be, however, a willingness to act on the best hypothesis that one has time or opportunity to form.

14. *Distinguishing between hypotheses and solutions.* The average person, once he is forced is think about a problem at all, soon gets an idea that he accepts immediately as a solution. It

would help along the progress of our culture immeasurably if we could teach whole generations of students to call this first idea a hypothesis, and to distinguish it very carefully form a solution. It is not a solution until it has been tried under experimental conditions and found to account for the observed facts more completely than will any alternative hypothesis.

15. *Awareness of assumptions.* There is an increasing disposition on the part of scientists to be aware that their investigations begin at a rather advanced point on a continuum of prior notions, which, for purposes of the investigation, must be regarded as assumptions. It helps a great deal to make these assumptions explicit and to reduce the necessary assumptions to the smallest possible number.

16. *Judgement of what is of fundamental and general significance.* Much of the work that goes on in scientific laboratories strikes the non-scientist as trivial. Imagine a man like Pavlov spending years of his life measuring how much a dog's mouth watered at the sound of a dinner bell ! But a scientist can first reduce a wide range of phenomena to a few crucial points that need to be investigated in order to explain how these phenomena work. Then, when he gets a few general principles out of his investigations, he sees how they will provide a fundamental, general explanation of all these diverse phenomena. The ability to dig down to basic causes is one mark of a great scientist and is probably rare. But we could find important differences among students in their ability to judge what evidence in an experiment was fundamental and conclusive and what other evidence was, or would be, irrelevant, superficial, or unimportant.

17. *Respect for theoretical structures.* The layman's idea of scientific work is that it consists of accumulating great numbers of facts from which general conclusions may later be drawn. But the facts are never accumulated at random; they are sought and sifted for evidence bearing on a theory. True, certain fact-gathering agencies like the Census Bureau accumulate many facts which they themselves do not propose in investigate, but only because these are basic data that have proved useful in many types of investigations. A scientist is not likely to adopt

the attitude expressed in the cliche : "That is all right in theory but won't work in practice." A scientific theory is "all right" only if it does work in practice.

18. *Respect for quantification.* What not all scientific work makes use of quantitative relationships, a surprisingly large fraction of it does. New mathematical techniques for analyzing and organizing data are constantly needed in scientific research, and scientifically inclined layman cannot afford to ignore this powerful tool in solving problems. Admitting in these times that one cannot understand mathematics is as shameful as admitting that one cannot read.

19. *Acceptance of probabilities.* Very few lines of investigation now yield yes-or-no, all-or-none answers. Instead, they yield statements of the probability of occurrence of an event, and whether this probability is greater than would be expected on the basis of pure chance. While it is less comfortable to depend on probabilities than certainties, modern men have to accept probability statements as useful and important, and appreciate the statistical laws governing their formulation.

20. *Acceptance of warranted generalizations.* The "general semantics" movement has created some distrust of general statements about classes of phenomena on the ground that each member of a class may have unique attributes that the generalization ignores. If pushed to its logical extreme, this position would deny the possibility of making any true statements about such a class as "lumber" on the ground that "Tree 1 is not Tree 2." While such caution may avoid certain errors, it must be recognized that general statements about classes of phenomena are the goal of science, not statements about particulars, and that such general statements based on research are necessary to enable the mind to grapple with the complexity of experience.

11

Scientific Attitude

Narendra Vaidya

The formation of scientific attitudes (temper) which is a processes that starts right from the very beginning in the immediate environment provide by the parents, friends, neighbourhood, school and society at large. Attitude is a 'condition of readiness for a certain type of activity". Attitudes held by the individuals may be simple or complex, stable or unstable, temporary or permanent and superficial or fundamental. Judgements based upon insufficient facts are likely to yield wrong results and thereby develop biased attitudes. Finding answers to problems through direct observation, adequate experimentation, argumentation on facts, verification and testing of knowledge are some of the initial manifestations of scientific attitude. Especially for young children, these manifestations need further to be concretized in terms of well planned teaching situations and meaningful activities. This built in intention in science teaching is more likely to result in the gradual growth of scientific temper among the vast body of school children. Some characteristics of this scientific temper are:

1. Open-mindedness.
2. Curiosity.
3. Judgement based upon scientific facts alone.
4. Willingness to test and verify conclusions.
5. Faith in cause and effect relationship.
6. Honest reporting.

7. Rejection of the principle of authority. And lastly,
8. More faith in the books written by specialists in their respective fields, etc.

The development of scientific attitudes should not be left to chance. Science teachers, on the other hand, should make a special effort to develop them. This they can do by employing democratic procedures in the classroom, engaging students on projects mostly of the investigative types, by emphasizing the need for the collection of data before conclusions are made and the choice of all these relevant activities in day to day teaching from significant areas of living which correct their erroneous ideas and misconceptions. Frequently, they must demand ample proof before they make firm judgements.

12

Scientific Attitude

M. S. Chhikara
&
S. Sharma

Teaching of sciences should aim at developing a particular bent of mind, in the students towards, ideas, events and living would that in based on scientific explanations. This is scientific attitude. For this purpose, this subject should be taught in a systematic way. Scientific attitude invokes several things but some major are:

(a) *Mathematical accuracy*: Students should observe mathematical accuracy in every type of collection and analysis of data.

(b) *Observation* : Students studying sciences should have a way of observing things critically. Their perception should have meaning and observations should be explained by applying the knowledge one has acquired.

(c) *Verification* : Students should try to verify the claims of other people and they should not accept those as such. One should not believe the tall claims of quacks about their medicine but should have a desire to test and verify those.

(d) *Impartiality* : Students should be impartial towards events and phenomena. Analysis and interpretation of data should be undertaken objectivily. Data should not be manipulated and results of experiments should not be influenced by personal prejudice or bias.

(e) *Experimentation* : Students should have a desire to undertake experiments and should not draw conclusions based on anticipations. In data to get accurate results controlled experiments should also be set up.

13
Scientific Attitude

Edward W. Smith
Stanely W. Krouse, Jr.
Mark M. Atkinson

The scientific attitude, by its very name, tends to be associated solely with the area of science and scientific methods. Actually, the scientific attitude is applicable to nearly every situation an individual may encounter in mathematics, social studies, science, and other subjects in the curriculum. It is closely allied to critical thinking, and it is developed through the study of language was well as other subject-matter areas.

The scic..ific attitude can be a valuable result of the problem-solving approach to learning. It is encouraged when study is attacked through (1) identifying a problem, (2) making valid observations (3) drawing objective conclusions, (4) verifying the conclusions, and (5) applying these conclusions to new but related problems. These steps are those ordinarily associated with scientific methods of investigation and closely allied to the steps involved in the problem-solving approach.

The solving of problems of in an orderly manner leads to the development of an objective attitude that is applicable to many situations. It can be developed in many subjects in the curriculum and applied in adult life as well as in school life.

The person with the Scientific Attitude

An individual who has learned the scientific attitude and makes use of it does not jump to conclusions. He is patient and reserved in his judgment. In considering a situation or a problem, he studies all aspects of it, looking at every side of it before approaching the study with a minimum of prejudice or bias. Although he may have previous knowledge, he will not have preconceived notions unless they have basis in his objective understanding of the problem.

The person who possesses the scientific attitude has no time for old wives' tales, rumor, or superstitions. He is person of caution who observes carefully before coming to a conclusions. He is ready and willing to change his mind when he observes new evidence that he can accept as valid.

The scientific attitude is reflected in the individual who :

1. Bases his actions, thoughts, and conducts on the best knowledge that is available to him.
2. Suspends reaching conclusions and forming judgments when reliable and objective information is lacking, or until such time as he has the opportunity to study such information.

Signposts of the Scientific Attitude

The development of the scientific attitude in students should be one of the goals of the school program. Its development in students leads to a willingness to work together for the solution of problem. It is conducive to the use of many sources for locating valid information. Critical thinking in evaluating and applying such information is also closely allied to the scientific attitude.

Students experienced in the use of the scientific attitude are willing to experiment, patient in checking the results of their experiments, and questioning as they conduct their experiments.

Individual without a proper attitude toward the application of knowledge or the acquisition of new knowledge gain little value from their education. Those with the scientific attitude have a means of applying their knowledge and a thirst for new knowledge.

Developing a Scientific Attitude

The Scientific attitude is not one that simply comes with maturity; it must be encouraged, practiced, and emphasized during the learning process. 'Careful planning is one of the most helpful ways of guiding students in the development of such an attitude. A course of study or a syllabus need not be confining, but such guides can offer many needed ideas, concepts, and facts that will aid the teacher and the pupil with their planing. If a course of study or other guide is lacking, cooperative planning can lead to areas that the class, guided by the teacher, wishes to pursue.

In any event, the teacher and the pupils may suggest topics and compile them, determining by a process of elimination those that will be considered. If all topics seem to be of equal importance, it might worthwhile to form committees to undertake different topics.

After the topics are selected, the class or the committees to which the topics are assigned should discuss suitable activities, demonstrations, or experiments that will help them develop the topic. Those to be used should be decided upon. After this is completed, lists should be made to include reference materials, community resources, equipment, and supplies that are needed. Arrangements for these resources of information should be made, and questions regarding each topic should be listed as a guide to the research into the topic.

After the questions have been decided upon, use should be made of the available resources. The reference books, the textbooks, and the community resources should be consulted in an effort to completely answer the questions. Additional questions that arise during the research should also be listed for discussion with other members of the entire group, who will determine whether they should also be investigated. Activities and experiments should be planned as to their purpose and the procedure to be followed. These should then be completed.

When the questions have been answered, the class or committee should come to agreement on the form of report to be presented, either orally or in writing.

It is then time for the preparation of the report, based on the information that has been compiled. The class will check to be certain that all of the questions have been answered and that the report is complete, including a listing of sources of information.

After the report is shared with all members of the class, it should be discussed. Sources of information, conclusions, facts, and background information contained in the report may be questioned. It is often helpful to have a guide sheet developed by the class to act as an agenda for such discussions. This would include (1) presentation of the report, (2) completeness, (3) accuracy, (4) originality (5) informational nature, (6) sources, and (7) conclusions.

One seventh-grade class developed certain points that they left should be considered as the basis for judging the reports that were presented to them. This class checked each question after reports were given and then offered suggestions to help in future reports. Their points involve not only the information and accuracy of the report, but also its presentation:

1. How well was the report of research work prepared ?
   a. Was it complete ?
   b. Was all the information important to the topic?
   c. Was it well planned ?
   d. Did we learn something from the report ?
   e. Did we learn something from the report ?
   f. Did the person preparing the report learn something ?
   g. Was the report given in the person's own words ?
   h. Did the person use many sources of information ?
   i. If the pictures were used, were they worthwhile ?
   j. Was the report given in an original way ?
2. How well was the report delivered ?
   a. Did the person speak loudly and clearly?
   b. Did the person stand straight ?
   c. Did the person speak to the whole class ?
   d. Was the report given in good order ?
   e. Did the person know what he was talking about and understand all the words he used ?
   f. Did he use good English ?

g. Did he have to read his report ?

h. Did he give us a chance to ask questions ?

i. Could be answer the questions we asked?

The above questions offer a basis for class discussion of the report itself as well as the learning that resulted from it. Following such a discussion, students will profit from a summary of the knowledge that resulted from both the preparation and presentation of the report.

Such procedures will help the growth of the scientific attitude because of (1) the careful planning, (2) attention to the topic at hand, (3) the use of many sources of information, (4) the evaluation of the sources, (5) the accuracy of the information, (6) the procedures used in obtaining information, and (7) the learning that resulted. The encouragement of these procedures with a stress on their importance in developing the scientific attitude will aid in its growth, for results will be minimal unless the teacher is aware of the goal and the students share in this understanding.

Applying the Scientific Attitude

Each day learning and living in the school efforts multitudes of opportunities for developing the scientific attitude in the school community. Elections of pupils to student governments in high schools give teachers an opportunity to consider with the youngsters the leadership qualities that are required by the offices to be filled. Although the student should make his own choice based on the qualifications needed and the candidates for the offices, such an experience can provide a valuable lesson in the need for good judgment in selecting leaders.

The planning of a program for foreign-exchange students in a high school allows students an opportunity to participate in charting a course of actions involving objective data, as they select a student who might be suitable to the community.

In such groups as honor societies, election of new members to the groups based upon a specific criteria indicate the importance of objective thinking and the exercise of the scientific attitude.

Fund-raising campaigns for worthy causes allow students as opportunity to enter into joint comparative planning.

The selection of a senior play, the casting of the play, and the final presentation are good training grounds for developing a scientific attitude involving objectivity, open-mindedness and willingness to alter performances to suit others.

Evidence of a Growing Scientific Attitude

The seventh-grader will undoubtedly show more evidence of the scientific attitude than the second-grader, but if all teacher pointedly work toward the development of such an attitude, it will be evident throughout the child's school years.

The first-grader who checks his teacher's statements at home may be less impertinent than curious.

The seventh-grader who takes time to prove his mathematics problems is certainly showing that he wishes to check his results carefully.

The desire to seek more information about a topic is evidence of caution and careful judgment.

Students who are able to cite the sources of their information in class discussion certainly have consulted references for information.

The young student who laughs at superstitions because he realizes that nothing happens without some cause has mastered more of the scientific attitudes than many adults.

The primary youngsters who bring in reptiles, frogs, rocks, leaves and other material that may appal the teacher is certainly curious. The teacher has a fine opportunity to lead him into careful observation and conclusions.

The boy or girl who hesitates in answering a question that involves drawing a conclusion from evidence may not be inattentive, but rather carefully considering the problem from different aspects.

The Scientific Attitude Is Developed

Although many of the characteristics that we associate with the scientific attitude may develop naturally with maturity, the open-minded, cautious, objective approach to situations that require decisions based on reliable evidence should be encouraged and emphasized by each teacher. With proper attention to its development, the scientific attitude can be fostered as characteristic valuable to each student as he faces problems in schools and in adult life.

14
Scientific Attitude

Pauline V. Young
Calvin F. Schmid

A scientific attitude is many things in many situations. It requires consistent thinking, stern pursuit of accurate data, stubborn determination to analyze one's own system of thinking and to take nothing for granted. Evidence, tests, proof are the pillars of stern court of "evidential confrontation."[1]

A scientific attitude is more than "objective," "dispassionate," "unbiased" devotion to collection and treatment of facts. To be sure a scientist avoids personal and vested interests. He deliberately looks for facts which may be death to casually formulated theory. He seeks facts which could substantiate theory and give facts new meaning and vitality. He does not tailor his views to fit preconceived notions or preferences of men in "chairs or the mighty." He is aware that strong emotion is notoriously inimical to clear thinking.

A scientific attitude is based on a complexity of elements. A. B. Wolfe stresses the fact that "it must be firmly borne in upon us that scientific attitude rests upon one, and only one, fundamental article of faith-faith in the university of cause and effect. Without this faith, a steady, undaunted pursuit of scientific knowledge as a guide to action may be incontinently flouted whenever it interferes with special interest or prejudices."[2] According to Wolfe, as to most scientists, causation to terms of impersonal, phenomenal correlation and sequence is the essence of scientific inquiry. In other word, science is deterministic. The researcher who does not become thoroughgoing determinist does not acquire a scientific attitude.

" 'Cause' is our name for the inter-connection among natural processes in their contingent relations," says G. Coghill, naturalist and philosopher. "Cause is inherent in the organization of the total situation. It is not imposed upon it from the outside." To answer the research "why" is to account for the facts by showing or identifying the conditions under which events take place.[3]

Many scientists believe that the problem of modern social science relates to casual explanation. A sound foundation in any science is laid through an understanding of the *processes of becoming*. Social science cannot avoid this task. "Social becoming, like natural becoming, must be analyzed into a plurality of facts, each of which represents a succession of cause and effect."[4]

The characteristics of a scientific attitude, discussed below, may sound like a big order even for a mature scientist. But he is not expected to fulfil this role at all times, not in its entirely at any time. The young student is not expected to acquire scientific attitudes at the start of research project. Many students, however, strive for and do consequently develop a scientific spirit and derive a genuine thrill in the pursuit of scientifically conducted research. One author compares this thrill in science to that resulting from finding a gold nugget after shoveling dirt in a gold field for a long time. "It is worth all the drudgery that preceded it," he adds.[5]

To Sir Francis Galton a scientific attitude meant "an inherent stimulus to climb the path that leads to knowledge with the strength to reach the summit— one which, if hindered and thwarted, will fret and strive until all hindrance is overcome, and it is again free to follow its labor-loving instinct."

A scientist may encounter many failures and delay in his search for data or in his attempt to confirm their validity and accuracy. Without losing courage and enthusiasm, he must go on undaunted and undismayed. The basis for success is discovery lies in cultivating an ability to rise from failure with a curiosity and new sense of inquiry.

To John Dewey a scientific attitude was a linked with "an ardent curiosity, fertile imagination, and love of experimental inquiry".[6] He pointed out that "young people who have been trained in all subjects to look for social bearings will also be educated to see the causes of present [problems and phenomena] they will be equipped from the sheer force of what they have learned to see the possibility and the needs of actualizing them. They will be indoctrinated in its deeper sense without having doctrines forced upon them."[7]

Herbert Blumer—in describing Thomas and Znaniecki as "scientists at work"— saw these characteristics in the eminent authors: "....... rich experience with human beings, keen sensitivity to the human element in conduct......lively curiosity.....[endless] mulling over [their data], reflecting on them perceiving many things in them, relating these things to their background and experience, checking these things against one another, and charting all of them into a coherent analytical pattern."[8]

To Allyn A. Young a *scientific attitude* means the ability to raise significant questions and to formulate fruitful hypotheses: "Successful research calls for industry and command of the appropriate technical methods. But if there is to be more than fact-finding, it calls for imagination, for ability to see a problem, and to devise hypotheses that are worth testing. Industry fortunately is not an uncommon virtue. Techniques may be acquired. But imagination, and especially the kind of imagination that keeps its moorings, is rare."[9]

Indeed, a whole large volume was recently devoted by C. Wright Mills[10] to the necessity of sociological imagination in the study of social life. According to Mills, sociological imagination is that quality of mind which aids students in using "information and developing reason in order to achieve lucid summation of what is going on in the world and of what may be happening within themselves." He urged fellow sociologists to cultivate in students an imagination not only as an aid in acquiring an orderly knowledge of science but also in coming to grips with the major problems of the day as well as with their personal problems.

Charles Horton Cooley, in characterizing the scientific attitude of his colleague, William Graham Sumner (author of the famous volume, *folkways*), said: "What distinguishes him and makes the manner of his work a possible source of help to others is something inseparable from his personality—his ardor, his penetration, his faith in social science, his almost incredible power of work, his great caution in maturing and testing ideas before publication." The characteristics of scientific attitude described here are not necessarily those of eminent scholars alone. Frequently we find these traits in young students who have become interested in a piece of research which has fired their imagination.

Sir Francis Bacon conceived a true scientist to processes both "compassion and understanding, since knowledge without charity could bite with the deadliness of a serpent's venom. And mere power and mere knowledge exalt human nature but do not bless it."[11] He urged the scientist to keep close to reality; to ascertain, in a dedicated manner, search for new discoveries. The development of the experimental method alone, Bacon had contended was the means by which all things else might be discovered.

In short, at present as in generations past, among mature scientists and among students, we find a compelling and what Karl Pearson called a "wonderful restlessness," a lively curiosity, endowed with imagination in the study of man. Social scientists in particular are compelled as if they were charged with a great responsibility, so acquire knowledge of human behaviour and social processes.

Notes :

1. Herman Wouk, *This Is My God*, p. 226. Doubleday & Co., 1959.
2. Wolfe, *Conservatism, Radicalism, and Scientific Attitude*, p. 203. The Macmillan Company, 1923.
3. Coghill, *World, Regions and Man*, p. 1. Cheshire Publishing House, 1960.
4. Thomas and Snaniecki, *The Polish Peasant in Europe and America*, p. 36. Knopf, 1927.
5. "The Young Scientist," *Fortune*, (June 1954), p. 182.
6. Dewey, *How We Think*, p. iii. D. C. Heath and Co., 1933.
7. Dewey, *Problems of Men*, p. 183. Philosophical Library, 1946.
8. Blumer, *A Critique of Research in the Social Sciences*, p. 76. Social Science Research Council, 1939.
9. Young, in Wilson Gee, *Social Science Research Methods*, p. 362. Appleton-Century-Crofts, 1950.

10. *Mills, The Sociological Imagination*. Oxford University Press, 1959.

11. Bacon, *The Advancement of Learning*, pp. 75–100. Oxford University Press, 1868.

15
Scientific Attitude

V. K. Kohli

One of the chief aims of science teaching, is to develop certain scientific disciplines or attitude. Before we can know the methods of inculcating such attitudes, let us first be clear about its meaning and concept.

Definition

According to NSSE, "Scientific attitudes can be defined as open-mindedness, a desire for accurate knowledge, confidence in procedures for seeking knowledge and the expectation that the solution of the problem will come through the use of verified knowledge".

According to another view, scientific attitudes include—freedom from bias, prejudice and superstitions, open-mindedness, critical mindedness, intellectual honesty, beliefs when new evidence is available.

Qualities of a person who possesses scientific attitudes

1. He looks for the natural causes for the thing that happen, i.e. he
   i) does not believe in superstitions, such as charms or good or bad luck,
   ii) believes that there is no connection necessarily between

two events just because they happen at the same time one after the other.

2. He is curious concerning the things he observes, i.e., he
   i) wants to know the 'whys', 'whats', 'hows' of observed phenomena,
   ii) is not satisfied with vague explanations of his questions.
3. He is open-minded towards works and opinions of others and information related to his problem, i.e. he
   i) believes that truth never changes, but his ideas of what is true, may change as he gains better understanding of that truth.
   ii) revises his opinions and conclusions in the light of additional reliable evidence.
   iii) listens to, observes, or reads evidence supporting ides contrary to his personal opinions.
   iv) accepts no conclusion as final or ultimate.
4. He evaluates techniques and procedures used and information obtained, i.e., he
   i) uses a planned procedures in solving his problems.
   ii) seeks to use the various techniques and procedures which have proved valuable in obtaining evidence.
   iii) seeks to adopt the various techniques and procedures to the problem at hand.
   iv) personally considers the evidence and decides whether it relates to the problem.
   v) infers whether the evidence is sound, sensible and complete enough to allow a conclusion to be drawn.
   vi) selects the most recent, authoritative and accurate evidence related to the problems.
5. His opinions and conclusions are based on adequate evidence i.e., he
   i) is slow to accept as facts anything not supported by

convincing proof.

ii) bases his conclusions upon evidence obtained from a variety of dependable sources.

iii) searches for the most satisfactory explanation of observed phenomenon.

iv) sticks to the facts and avoids exaggeration.

v) does not allow his personal pride, bias, prejudice or ambition to change the truth.

vi) does not jump to conclusions.

Views of NCERT

The National Council of Educational Research and Training (NCERT) conducted a workshop at Chandigarh in 1971 and evolved the following specific behaviour of a pupil who has developed scientific attitude:

The pupil—

1. is clear and precise in his statements and activities.
2. bases his judgement on verified facts (not on opinion).
3. is willing to consider new ideas and discoveries (free from prejudices).
4. reacts favourably to efforts made to use science towards human welfare.
5. is prepared to reconsider his own judgements.
6. arranges the apparatus, materials etc., in their proper places at the end of the work.
7. suspends judgement in the absence of sufficient data.
8. is free from superstitions.
9. is objective is his approach.
10. is honest and truthful in recording and collecting scientific data.

Techniques for Developing Scientific Attitudes

As discussed earlier there is clear cut definition of scientific attitudes. Different people opine differently. This is why there are no exact and concrete ways of developing these attitudes.

Scientific attitudes are certain mind-sets in a particular direction. So by adopting varied techniques, such mind-sets may be created. According to one view, direct teaching does not produce important changes in pupil's attitudes, while on the other hand, out of school uncontrolled experiences are most important for developing these attitudes. Curtis shows by his investigation that direct teaching does modify the attitudes of young people. Tvler proposed the following suggestions for planning learning experiences to inculcate scientific attitudes :

1. Increase the degree of consistency of the environment.
2. Increase the opportunities for making satisfying adjustments to attitude situations.
3. Provide opportunity for the analysis of problem or situation so that a pupil may understand and then reset intellectually in the desirable attitude.

Realistically speaking, major responsibility for developing scientific attitudes among students lies on the science teacher. He can manipulate various situations to infuse among the pupils certain characteristics of scientific attitudes. Moreover he can give his own practical examples by possessing and practising various elements of these attitudes. This way of copying the teacher will leave a permanent work on the personality of the students.

However, apart from above, following points are worth trying for the teacher to develop scientific attitudes:

1. Use of Wide Reading

According to a study made by Curtis, the pupils who engage themselves in wide reading in science, develop scientific attitudes more than those who study only one text book. It indicates that students should be encouraged to read library books and supplementary books on science. It will be possible only if every school has a Science Corner in the school library or preferably a separate Science Library and the science teacher himself has a love for extra-reading. He should then transfer in the pupils this love for reading and inculcate the ability to use and understand reference books. The teachers should be a person who grows professionally, reads new title and does not come to the end of his subject but shares his joys of new reading with his pupils and refers to them certain suitable books, for in the words of Dr. Rabinder Nath Tagore:

> *"A teacher can never truly teach unless he is still learning himself. A lamp can never light another lamp unless it continues to burn its own flame. The teacher who has come to the end of his subject, who has no living traffic with his knowledge, but merely repeats his lesson to his students, can only load their minds. He cannot quicken them."*

2. Study of Superstitions

Just by taking about superstitions and unfounded beliefs in the class and calling them bad and out of date, will not leave any impression on the minds of the pupils. When you teach so many science lessons practically, it will be more effective if the science teacher encourages his pupils to practically investigate some common superstitions and come to their own conclusions by actual survey and study. For example, one popular superstitions is that if there is a broken mirror in a home, there can not be peace and harmony in the family. This thing can be studied by deliberately possessing a broken mirror in the school or some students may have in their homes while others should keep unbroken mirrors. For a week or fortnight a comparative study can be easily made and the pupils can come to their own conclusions.

In Northern India there is another popular belief that if early in the morning, when you first step out of your home and come across a cat before meeting any other individual, you are doomed, i.e., the day is a bad day for you and you won't see any of your work done that day. This belief can also be studied and investigated in the similar manner by actually providing situations and the students can arrive at their own results.

Research on the subject has shown that by encouraging practical survey and study of such common beliefs, students have developed permanent mind-sets or attitudes towards such superstitions.

3. Use of Planned Exercises

Some magazines devoted to science provide exercises for

developing certain attitudes. Proper use of such exercises should be made quite frequently. Cutting from new-papers can also be used for the purpose. Certain pictures and cuttings may be displayed on the bulletin board and used again and again for direct teaching.

Good text-books contain exercises at the end of each chapter which fulfil the aim of developing scientific attitudes. For example, questions of the style given below, are to be answered by the pupils.

"Some people have this belief that if a crow happens to fly into à home through and open door or window, then some death is sure to occur in the home. Which of the following statements, do you think,, is most appropriate ?

i) The belief is nonsense.

ii) There is some doubt that the belief may be true.

iii) Perhaps there is no possibility of any truth of statement.

iv) For some people the belief may be true.

v) It can be true under certain circumstances.

4. Proper use of Laboratory Period

The laboratory period can offer many opportunities for learning certain elements of scientific attitudes. It is the function of the teacher to see that such opportunities are properly utilized. He should see that the problem of the experiment is clearly stated. The hypotheses on the solution are present and proper method of testing the hypotheses is practised. After the experiment, the result should be discussed and interpreted. The pupils should be taught to suspend judgement if sufficient evidence is not forthcoming.

5. Co-curricular Activities in Science

A host of such activities like Science Club, Hobbies Club, Scientific Society, Chemical Society, Photographic Club, Scientific Excursions, Making of Scientific Models, Organizing Science Fairs and Science Exhibitions, Improvising Science Apparatus, Maintaining Aquarium, Vivarium etc., should be effectively organised and the pupil be given sufficient freedom to plan their activities. Indirectly the student will be inculcating some desirable attitudes.

6. The Atmosphere of the class

If the Internal setting of the class is properly arranged and the room is decorated in a manner which contributes to the development of proper atmosphere, then to some extent, the thinking of the pupils can be diversed towards the inculcation of certain attitudes.

Secondly, the role of the teacher is also very important to develop desirable atmosphere in the class. The teacher should encourage the spite of friendly criticism of procedures, data collection, hypotheses and results. While teaching in the class, he should see that the lessons are charged with intelligent questions of students. It has been commonly see that very many teachers snub such pupils who ask more questions. This attitude of the teacher will not help to develop proper atmosphere in the class.

7. The personal Example of the Teacher

Perhaps the greatest force for the inculcation of scientific attitudes is the personal example of the teacher. The knowledge of Psychology tells us that the children have a great tendency to copy the teachers. To much extent it can be said, "As is the teacher, so is the student." Therefore the science teacher himself must be free from bias and prejudices while dealing with the pupils. He must be open-minded, critical in thought and action in his day to day dealings, must be free from superstitions and unfounded beliefs, must be objectives and impartial and his approach to everyday problems must be truthful and having faith in cause and effect relationship.

16

Learning Scientific Attitudes

Nelson B. Henry

Views as to the Nature of Scientific Attitudes and Methods

In the present century scientific methods and attitudes have been emphasized, particularly by Dewey, as objectives of formal education.[1] Many articles and investigations on the nature of scientific method have been published since Dewey's analysis. An inspection of this literature leads one to assume that such phrases as "scientific methods," "methods of science," "scientific thinking," "problem-solving," and "critical thinking" mean much the same and may be used interchangeably without undue confusion.

There developed through the years an interpretation that there was one scientific method with define steps to be followed in a sequential order. Conant[2] and others have pointed out that this is not an acceptable interpretation. At present, most writers on this subject feel that there are many ways to solve problems scientifically and that no one way is necessarily typical of scientific thinking. It is also to be noted that the methods of science do not necessarily produce a successful solution to every problem.

Some writers on scientific method emphasize the inductive process; others the deductive process. However, it seems reasonable to infer that both processes are involved, at least in certain kinds of problems. It has been argued as to whether scientific methods are largely empirical or largely theoretical. There appear to be learning activities that are mainly empirical and descriptive, others that are highly conceptual, and still others in which theory and trial and error are both employed.

Some Representative Scientific Attitudes

The first comprehensive study to be published with respect to scientific attitudes was Curtis's[3] analysis. Noll's[4] list included such habits as those of intellectual honesty, open-mindedness, and suspended judgment. Other investigators have cited such manifestations as freedom from bias, looking for cause-and-effect relationships, curiosity, and criticalness. Extensive lists of scientific attitudes are to be found in the investigation of Crowell,[5] Ebel,[6] and Lampkin.[7]

Components of Scientific Methods

As indicated previously, several different phrases have been employed to designate the methods of science. It appears from an examination of the literature that such phrases refers to some common components. The components which seem to be most frequently mentions are: recognition of problem, collection of relevant data, formulation of hypothesis, testing of hypothesis, and drawing conclusions.

The interested investigator should consult the works of Downing, Crowell, Keeslar, and Lampkin[8] for detailed treatments of the methods of science. The individual interested in setting classroom situations for the direct teaching of the attitudes and methods of science and for the testing of such attributes should also find the writings[9] of Branard, Obourn, and Burmester worth while.

Research studies have indicated that the acquisition of scientific methods and attitude is facilitated by instruction. The evidence as to the relative effectiveness of direct as opposed to indirect teaching favors the former. Studies, with some notable exceptions, show that although there is a positive correlation between the amount of factual knowledge acquired through science training and the ability to exhibit some scientific attitudes,[10] teaching matter alone does not produce significant changes in attitude nor measurably train in scientific method.[11] A few thoughtful studies to the contrary, the evidence,[12] on the whole, is convincing that direct teaching for scientific attitude and of scientific method is more effective than instruction not focused directly upon these outcomes.[13]

Experiences Conducive to the Learning of Scientific Attitudes and Methods

The principles of leaning which are to be observed in teaching directly for the attitudes and methods of science are the same as those applicable for any other educational objective. The experiences should be psychologically sound, with due cognizance given to student aims and needs. There should be student activity— such as would be in agreement with the type of learning involved in the student's objective. There is also need for wise direction for the student's endeavors. The teacher's own attitudes and methods are certain to be influential in such learning situations.

With reference to the acquisition of scientific methods attitudes, it seems obvious that if students are to develop these abilities they must have practice in them. That is, situations should be designed to allow students to select worth-while problems and attempt to solve them. They should have experiences in collecting data, making guesses, devising experiments, and checking for accuracy while cultivating methods and attitudes conducive to effective learning in the field of science. Suggestion for analyzing and carrying out such operations have been made by Goldstein.[14]

EVALUATION OF INDIVIDUAL APTITUDE IN SCIENCE

The Distinction between Aptitude and Achievement

The major distinction between aptitude and achievement is temporal: achievement refers to present or past accomplishments; aptitude relates to the possibility of future accomplishment. Achievements depend, in part, or prior aptitudes, and aptitude as a predictor of future achievement involves present and past accomplishment. Despite this ambiguity in expression, the terms here employed have some utility.

For persons untrained in a field, aptitude, either on logical or empirical grounds, may be found to be a complex of intelligence, reading ability, reasoning skills, interests, attitudes, and motor skills. At a more advanced level, such as for graduate work in science achievement in undergraduate courses as determined by grades or tests will become an additional and significant component of aptitude. As these last two sentences imply, attempts to assess aptitude may create a wide array of relevant concepts such as knowledge, intelligence, personality, physique, and motor skill. Achievement, on the other hand, refers to tangible accomplishments in the field which in science might be exemplified by knowledge, laboratory skills, or completed research.

The Components of Scientific Aptitude

The phrase "scientific aptitude" appears to have clear meaning until the attempt is made to define it. The simple phrase, then, involves a complex of interacting hereditary and environmental determiners which produce the predispositions or abilities spoken of as scientific aptitude. All studies indicate that high intelligence is very significant for success in any task dependent on the manipulation of symbols and the formulation of concepts and conceptual schemes (theories). High intelligence is essential to scientific achievement. Additional mental factors that appear to be associated with success in science are intellectual curiosity, ability to apply knowledge to new situations, retentive memory, and insight into abstractions.[15] These attributes are similar to those found in generally gifted individuals. Such factors as physical development, social and emotional maturity, moral character, interests, attitudes, and skills may also be facets of scientific aptitude.

Relation of Scientific Aptitude to Attitudes, Critical Thinking, and Other Qualities. The use of the phrase scientific aptitude implies that persons possessing certain characteristics can be identified and that such individuals can succeed in scientific endeavour. Thus, the characteristics of able scientists suggest some of the criteria for locating individuals with aptitude for science. These characteristics include mental acuity, creative abilities, capacity for critical thinking, ability to see relationships, suspended judgment, and open-mindedness. Factors that predispose to such trains constitute at least a part of scientific aptitude.

Patterns of Ability and Predictions of Success. Brandwein postulates three factors.

1. A genetic factor for verbal and mathematical potential.
2. A predisposing factor consisting of persistence and "questing."
3. An activating factor opportunities and a special kind of teaching.[16]

MacCurdy[17] reported detailed descriptions of the superior science student with respect to personality, attitudes and opinions, activities, interests, family history, associates, science teacher, and decision to be a scientist. Some of the traits cited were : leadership, self-control and self-discipline, curiosity, persistence, antisocial attitudes, excellent performance in scholarly activities, and manifestation of scientific interest. He concluded that capacity, interest, and freedom are qualities that are paramount in the life of the superior science student. Neivert[18] concluded that high intelligence, opportunity for development, and personal attributes were the three factors necessary for high science potential and for determining the selection of science as a career. She listed the science teacher as the most influential single factor in the school environment for the development of the potential scientist. By realising him from routine requirements and replacing these by more challenging opportunities, and by supplying or suggesting sources of materials and equipment, the science teacher may succeed in stimulating the able student.

Patterns of ability in science seem to follow the general patterns that have been cited for the gifted child. Such a child tends to be taller, healthier, and stronger than other children and relatively free from nervous disorders.[19] The gifted pupil also tends to show superiority in desirable personal traits, to possess superior self-criticism, to receive more opportunities as a leader, and to stand above the average in moral qualities.[20]

Success in college is a prerequisite to success in science in the present day. There have been many studies of the factors involved in success in college. Burnett says that studies of student success in college can best be predicted on the basis of intelligence and reading comprehensions, together with high-school achievement indicative of self-sufficiency that might be epitomized as self-responsibility and study ability. He states further that achievement scores in specific subject provide the lowest predictive correlations.[21]

References :

1. John Dewey, *How We Think*, Boston : D.C. Heath & Co., 1933 (revised).
2. James B. Conant, *On Understanding Science*, New Haven, Connecticut : Yale University Press, 1947.

3. Francis Day Curtis, *Some Values Derived from Extensive Reading of General Science*. Contributions to Education, No. 163. New York : Bureau of Publications, Teachers College, Columbia University, 1924.

4. Victor H. Noll, "Measuring the Scientific Attitude," *Journal of Abnormal and Social Psychology*, XXX (July–September, 1935), 148.

5. Victor Crowell, "The Scientific Method," *School Science and Mathematics*, XXXVII (May, 1937), 525–31.

6. Robert L. Ebel, "What Is the Scientific Attitude ?" *Science Education*, XXII (February, 1938), 75–81.

7. R. H. Lampkin, "*Scientific Attitudes*," Science Education, XXII (December, 1938), 353–57.

8. Elliott R. Downing, "The Elements and Safeguards of Scientific Thinking," *Scientific Monthly*, XXXVI (March, 1928), 213–43; Crowell, *op. cit.*; Keeslar, *op. cit.*; and Lampkin, *op. cit.*

9. Darrell Barnard, "The Lecture-Demonstration versus the Problem-solving Method of Teaching a College Science Course," *Science Education*, XXVI (October–November, 1942), 121–32; Ellsworth S. Obourn, "An Analysis and Check List on the Problem-solving Objective," *Science Education*, XL (December, 1956), 388–92; Mary Alice Burnester, "Behavior Involved in the Critical Aspects of Scientific Thinking," *Science Education*, XXXVI (December, 1952), 259–63.

10. Sam Strauss, "Some Results for the Test of Scientific Thinking," *Science Education*, XVI (December, 1931) 80–93; Otis W. Caldwell and Gerhard E. Lundeen, "Students' Attitudes regarding unfounded Beliefs," *Science Education*, XV (May, 1931), 246–66; P. P. DeWitt, "Attitudes Related to the Study of College Science," *School Science and Mathematics*, XXXIX (June, 1939), 552–57.

11. Rosalind M. Zapf, "Suggestions of Junior High School Pupils," Part II, "Effect of Instruction on Superstitious Beliefs," *Journal of Educational Research*, XXXI (March, 1938), 481–96; George Wessell, "Measuring the Contribution of Ninth-Grade General Science Courses to the Development of Scientific Attitudes," *Science Education*, XXV (November, 1941), 336–39.

12. J. Wesley Eberhard and George W. Hunter, "The Scientific Attitude as Related to the Teaching of General Science," *Science Education*, XXIV (October, 1940), 275–81; James S. Perlman, "An Historical versus Contemporary Problem Use of the College Physical Science Laboratory for General Education," *Journal of Experimental Education*, XXI (March, 1953), 251–57.

13. Curtis, *op. cit.*

14. Philip Goldstein, *How To Do an Experiment*, New York : Harcourt Brace & Co., 1957.

15. *Education for the Talented in Mathematics and Science*, U.S. Department of Health, Education, and Welfare, Bulletin 1952, No. 15. Washington : Government Printing Office, 1953.

16. Paul F. Brandwein, *The Gifted Student as Future Scientist*, New York : Harcourt, Brace & Co., 1955.

17. Robert Douglass Mac Curdy, "Characteristics of Superior Science Students and Their Own Subgroups," *Science Education*, XL (February, 1956), 3 ff.

18. Sylvia S. Neivert, "Identification of Students with Science Potential." Unpublished Doctor's dissertation, Teachers College, Columbia University, 1955.

19. Marian Scheifele, *The Gifted Child in the Regular Classroom*, New York : Bureau of Publications, Columbia University, 1953.

20. *Education for the Talented in Mathematics and Science*, *op. cit.*

21. R. Will Burnett, *Teaching Science in the Secondary School*, New York : Rinehart & Co., 1957.

17
Development of Scientific Attitude

R. C. Das

The development of scientific attitude of mind is one of the objective of teaching science. It is very significant outcome of the process of science education. Teaching of science should not only enable the learners to master the facts, concepts and principles of science or develop instrumental and problem-solving skills but also develop scientific attitude of mind as well as interest and appreciation in them. Scientific attitude of mind is essential to enable them to adjust themselves and live as efficient citizen in a scientific society. The National Science Teachers Association of USA says that as a result of science education, the learners should be in the "process of developing a personal philosophy based on truth, understanding and logic rather than one based on superstitions, intuition or wishful thinking."

A scientific attitude can be developed only through personal experience and keen observation in the process of science learning. The teacher will have to provide situations in the classroom of field environment where the students can experience, see and feel the need for developing this attitude. For instance, open mindedness of the learners is necessary in scientific pursuits. They should respect others' opinion but at the same time believe only in verified facts. They should learn to observe and think critically and accurately. Accuracy and precision are essential in scientific experimentation. The purpose of scientific pursuit is to find the truth. There is no place for bias or prejudice if truth is to be revealed. The student's observation, therefore, should be unbiased and objective. Intellectual honesty is indispensable in the study of science. While solving a problem a scientist proceeds carefully and patiently, examines each step logically and holds back judgement until he is satisfied with the proof. These characteristics of any scientific pursuit should become a habit in the students learning science so that these are developed as a mental attitude in them. The students of science must never believe in superstition or hearsay. They rely in cause and effect relationship and verified facts or proof.

In this connection, the *Rethinking Science Education* mentioned the characteristics of scientific attitude as open mindedness, a desire for accurate knowledge, confidence in procedures for seeking knowledge and the expectation that the solution of the problem will come through the use of verified knowledge." These attributes of the mind are essential for solving a problem scientifically, be it a problem in the area of science or a social problem. It is true that the teacher will have to provide activities and situations where the students get an opportunity to develop scientific attitude. The learners may, in the process of studying science, miss these additional aspects involved in a scientific pursuit. The teacher should point them out to the students. Further, one textbook of science may not be successful in projecting various components of scientific attitude. The teacher should therefore encourage students to read different books on science because research studies have revealed that the students engaging themselves in wider reading inculcate scientific attitude more than those who read one textbook. But students should perform experiments or do projects in science themselves because without practical work in science, components of scientific attitude will remain unachieve by the students. Their personal experience is more valuable than verbal statements of scientific attitude. The development of scientific method and scientific attitude are constituents of the goals of general education and must strive to attain them through the teaching of science subjects.

A discussion on the important of honess doubt in science will be relevant here. It is sometimes seen that when a student doubts a statement of the teacher and puts questions to the teacher, he is usually silenced or even rebuked for not accepting the statement of the teacher. This practice is harmful; it thwarts the spirit of enquiry and honest doubt which, in fact, should be developed in the students. Moreover, this is not democratic teaching; it is authoritarianism. When the student is not convinced of some statement hesitates to accept it, he is considered sceptical by the teacher and he discourages him. Such authoritarian teaching retards the development of critical thinking and objective judgement, so important for scientific investigation. Such teaching imprints fixed ideas on the minds of the students. In science learning it is not wise to accept things blindly and without questioning if doubt arises. But unfortunately, in practice, it appears, the questioning student is often branded disobedient or aggressive. This is simply punished original thinkers and rewarding unprotesting students.

There are many instances in the history of science where scepticism lead to great breakthroughs. Healthy criticism and a sceptical attitude are considered essential ingredients in science by many authors of science education, because without a questioning mind, science will loose its very foundation, that is, dynamism and progressive character. Philip Abelson in the "The need for scepticism' has pointed out (*Science*, 1962) that the great shortage in science now is not opportunity, manpower, money or laboratory space. What is really need is more of the healthy scepticism which generates the key idea—the liberating concept." The teachers should always remember that without a questioning mind and a spirit of enquiry, studies in science will only mean acceptance of dogma and will never lead to development of scientific attitude in the learners. The students should not merely be supplied with information about science; they should be made to practice and observation science so that they get the opportunity to feel and develop the components of scientific attitude in their minds.

18

Techniques for Developing Scientific Attitudes

Eldwood D. Heiss
Ellsworth S. Obourn
Charles W. Hoffman

Problems solving in all of its elements is closely associated with a group of attitudes or mind sets which are important as outcomes of instruction in science. Some of the techniques that have been used for developing scientific attitudes will be reviewed.

What the research shows. The studies of Caldwell and Lundeen[1] seem to reveal that pupils in the junior high school possess unfounded beliefs about natural phenomena to a marked degree and, most important, that these beliefs influence the behavior reactions of the young people.

Carrying on independent investigation Wessell[2] and Lichternstein[3] found that direct teaching failed to produce important changes in pupils attitudes. Each of these investigators suggests that uncontrolled out-of-school experience and such factors as maturation, are very important in the development of the attitudes held by young people.

A pioneer study by Curtis[4] and more recent investigations by Blair and Goodson[5] and by Vicklund[6] seem to show that direct teaching does modify the attitudes of young people. The fact that there is some conflict in the present evidence concerning teaching scientific attitudes in science should lead us to use the best known methods in the classroom until further research information becomes available.

When we study the behavior of people at large, and note the extent to which action is based on emotional rather than upon reasoned judgment, it seems safe to assume that whatever can be done to associate behavior with desirable attitude in science teaching is of value. A few techniques that have been used will be reviewed.

The use of wide reading. The study made by Curtis[7] and cited above, gave rather clear evidence that pupils who engage in wide reading in general science develop scientific attitudes more than those who study only a single text. This would seem to indicate that whenever it is possible, the teacher should supplement the text books with reading on the problem under consideration. Much of the content of science is intimately bound up with fascinating historical and biographical incidents which make excellent material for supplementary readings and reports. Many of these incidents relate how the masters of science worked and what attitudes motivated their action. These should become a part of every course in science. If the references are not available for general reading the teacher might use a period for an inspirational talk of the life of some scientist concerned with the problem being studied.

The study of superstitions and unfounded beliefs. We assume that we live in an enlightened age free from the influence of superstitions. Pupils may be shown the fallacy of his assumption by making a survey of the magazines on astrology found on the newsstands or the horoscopes published in many widely read newspapers and on sale at drug stores. Time spent in analyzing horoscopes and articles in magazines on astrology will prove interesting and perhaps valuable in acquainting pupils with the problem.

Some teachers have found pupils very interested in making a survey of superstitions and unfounded beliefs. Such a study will provide a great deal of material for interesting class discussion. Just talking about scientific attitudes may not change the behavior pattern of young people. In some cases teachers have brought a ladder into the classroom and had an activity to see how many pupils would walk under it. Another suggestion is to have each pupil bring a small mirror and see how many will actually break it before the class. While these may have the appearance of stunts they do have elements of a direct attack upon the problem of attitudes.

The use of planned exercises. Newspapers and magazines are excellent sources of materials that may be collected and used for building planned exercises on attitudes. In one such file kept for a number of years by the writer may be found pictures of people walking under a ladder, the advertisement of an oil company showing an oil prospector using a divining rod, and many other such things. This material is used over and over both for bulletin board and for direct teaching.

Some textbooks in general science and biology provide exercises devoted to practice in the use of desirable attitudes. The following exercise is illustrative:

Some people believe that if a bird happens to fly into the house through an open door or window that a death is certain to occur in the family unless something is done to thwart the superstitions influence. Which of the following statements do you think would most nearly represent the reactions of a person who has scientific training ?

(a) There is probably no foundation for the belief.

(b) For some people the belief is probably well founded.

(c) The belief is silly.

(d) There can be little doubt but that be belief is well founded.

(e) While I do not believe in this, yet I am disturbed when a bird files into the house.

Following is a lesson plan that has been used by a teacher of general science in the direct attack upon one of the scientific attitudes.[8]

Purpose of the lesson was to teach a scientific attitude: A scientist is open-minded and waits for further evidence before he formulates and definite conclusion.

Teacher directed activities	*Pupil responses*
1. Of what larger unit is the earth a part ?	1. The solar system.
2. What are the various bodies of the solar system ?	2. The sun, planets, satellites, meteors, and meteorites.
3. Howe are these bodies arranged in the solar system ?	3. The sun is the center of the solar system. The planets revolve around the sun and are arranged in the following order : Mercury, Venus, Earth, Mars, Jupiter, Saturn, Uranus, Neptune, and Pluto.
4. Why is the sun found at the center of the solar system ?	4. The sun was the first body and the planets were originally a part of it.
5. How do the planets move around the sun ?	5. The diagrams show that they move in the same direction.
6. How do the satellites move ?	6. They revolve about the planets.
7. How do the planets vary in the number of satellites which revolve about them ?	7. The earth has one, Mars, two, Jupiter and Saturn many, Venus and Mercury none. The larger planets have many and the smaller ones few or none.
8. Have you ever wondered why the planets move as they do ? Why the sun remains at the center of the solar system ? Why the other bodies move as they do ?	
9. What group of workers can give us the best answers to these questions ?	9. The astronomers.
10. What problem would the astronomer have to find out before he could arrive at any answer ?	10. 1) He would have to find out how the solar system was formed.

Contd.

Teacher directed activities	*Pupil responses*
	2) He would have to find out why the bodies move as they do and why they are found at certain distances from the sun.
11. What method would the astronomer use in securing information that can answer the problem— How was the solar system formed?	11. He would (1) set up a problem; (2) gather information by studying the heavens through the telescopes; (3) arrange and record his data; (4) form a tentative conclusion, summary, or hypothesis; (5) formulate a tested conclusion.
12. Though the above method the astronomers have given the world several explanations or hypotheses. On the following mimeographed sheets several hypotheses are stated, together with the evidence used as a basis for the explanations. Just how far has the astronomer answered the problem ?	12. He has only formulated a tentative answer.
13. How can he test his hypotheses?	13. He can check it by further observations and by the opinions of other scientists.
14. What is the problem which you should keep in mind as you read? As you read, decide whether you will accept any one or all of the hypotheses.	14. How was the solar system formed? (Period of reading.)
15. Which of the hypotheses will you accept ?	15. I would accept the planetesimal hypothesis because it seems to have the best evidence. Chamberlain and Moulton succeeded in obtaining photographs of the nebulae which may show how the solar system was formed.

Contd.

Teacher directed activities	*Pupil responses*
16. Why was the nebular hypothesis accepted for nearly a century and a half by thinking individuals ?	16. Because it was the only explanation which was based on evidence that was available at that time.
17. Why was it later disproven ?	17. When the telescopes were perfected many of La Place's ideas were found to be incorrect.
18. If the nebular hypothesis were the only explanation which we would have today would you accept it ? Why ?	18. Yes, if there wasn't any other and it seemed reasonable. Because it is based on data and comes from an expert in the field.
19. How long would you accept or hold the hypothesis ?	19. Until a better explanation is given which is based on accurate data.
20. Why wouldn't you accept Jeans and Jeffrye's hypothesis ?	20. Because there isn't enough evidence to support it, it is so new.
21. Would you reject it entirely ?	21. No, because the men are trained scientists in their field.
22. Would it be possible for Jeans and Jeffrey's hypothesis to replace the planetesimal hypothesis ?	22. Yes, if they succeed in obtaining better and more data.
23. What should be our attitude toward tentative explanations which the scientists give us ?	23. We would accept it until we have information which shows that it is incorrect.
24. What important idea has the lesson taught you ?	24. A hypothesis should be accepted until sufficient information is available.
25. In what everyday situations that you can mention would it be important for you not to jump to conclusions but maintain an open mind until you have proof ?	25. 1) In answering or arriving at conclusions to problems in the unit. 2) In deciding whether we liked certain people or certain things. 3) Our likes and dislikes may be biased by what others say.

The use of the laboratory period for developing desirable attitudes. The laboratory period in science offers many opportunities for practicing good attitudes. The teacher should see to it that the problem of the experiment is clearly stated, that hypotheses on the solution are presented, and planed for testing the hypotheses on the solution are presented, and plans for testing the hypotheses are made. In this discussion the matter of control and experimental factors should be brought out. After the experiment has been performed the evidence should be discussed and interpreted. The pupils should decide whether there is sufficient evidence upon which to base a conclusion or if more evidence is needed. They must be taught to suspend judgment where this is necessary.

The influence of the teacher and the atmosphere of the classroom in developing desirable attitudes. Perhaps the greatest force at present in the development of desirable attitude is a teacher who practice them day after day in his classroom. The atmosphere of such a classroom will be charged with a spirit of friendly criticism of procedures, data, hypotheses, and conclusions. It will encourage an intelligent questioning of authority and maintain a skepticism for reported evidence. Emotional and wishful thinking will be questioned, and prejudice and intolerance will find no place. Facts and assumption will be clearly distinguished, and hypotheses modified in the light of new evidence. If such an atmospheres modified in the light of new evidence. If such an atmosphere could be maintained in our science classrooms we could go a long way toward inculcating desirable attitudes even without the benefit of test results.

References :

1. Caldwell, Otis W., and Lundeen, Gerhard, "A Summary of Investigations Regarding Superstitions and Unfounded Beliefs." *Science Education* XX Frebruary 1936.

2. Wesseli, George, "Measuring the Contribution of the Ninth-Grade General Science Course to the Development of Scientific Attitudes." *Science Education*, XXV, November, 1941.

3. Lichtenstein, Arthur, "The Effect of Teaching Stress Upon an Attitude." *Science Education* XIX, April, 1935.

4. Curtis, F.D., "Some Values Derived from Extensive Reading in General Science." Teachers College, Columbia University, New York, 1924.

5. Blair, Glenn M., and Goodson, Max R., "Development of Scientific Thinking." *School Review* XLVII, November, 1939.
6. Vicklund, C.U., "The Elimination of Superstitions in Junior High School Science." *Science Education*, XXIV, February, 1940.
7. Curtis, F.D., *loc, cit.*
8. This lesson plan was prepared by Miss Edith Selberg of the Colorado, State College of Education, Greeley, Col.

19

Development of Scientific Attitude

J. K. Sood

It is expected that the study of biology should develop a scientific attitude. The development of the scientific attitude is possible only through conscious attempts to make it happen. To achieve this you should understand what it means. The scientific attitude is an attitude which reflects scientific thinking. And to be scientific means that one has such attitudes as curiosity, rationality, suspended judgement, open mindedness, objectivity and humility. Such characteristics of a scientifically minded person can also be called as components of the scientific attitude. A brief discussion is given below:

i) **Curiosity** is the first component. This is the desire for understanding new situations. A curious person asks questions, reads to find information and readily initiates investigation. Curiosity is a stimulus to inquiry.

ii) **Willingness to suspend judgement** is another component. People collect sufficient evidences before drawing conclusions or making judgements. They are ready to change their opinion in the light of new evidences.

iii) **No faith in superstitions.** A person with scientific attitude realizes that nothing happens without some cause. He is not open to false beliefs.

iv) **Open mindedness** is very close to suspended judgement. People in general and scientists in particular, are ready to change their opinion if sufficient evidence is available. Generally, one tries to feel attached to the old beliefs or ideas, but changing opinion on sound evidences is just to learn something new and novel.

v) **Objectivity** is also an important component of scientific attitude. All scientists collect and interpret data. It is expected that the interpretation will be done objectively and proper conclusions will be drawn based on data.

vi) **Intellectual honesty** is concerned with the conscious act of truthfully reporting observations. People should be intellectually honest in presenting their views or debating issues.

vii) **Liking for scientific explanation** gives a chance to a person to view things objectively or empirically. One should not depend completely on beliefs and mythological explanations.

viii) **Desire for completeness for knowledge** is a personality trait of all scientists and other intellectuals. People try to see the things in complete form and try to fit every piece into its place.

Similarly, there are other components of the scientific attitude such as judgement of what is fundamental and general, respect for theoretical structure, acceptance of probabilities and humility.

To foster the scientific attitude through biology teaching, it is essential to take some measures. The following points will prove helpful in achieving it :

i) the particular component of the scientific attitude, which is to be developed, should be identified.

ii) The meaning or the behaviour related to the particular component should be explained.

iii) Learning experiences, which will be helpful in developing the scientific attitude, should be given to the learner. The pupil should participate effectively in such activities.

iv) Along with learning experiences, some emotional attachment should also be developed so that the pupil may feel attachment with a particular idea.

v) Appropriate situations should be provided for practice and confirmation.

vi) Deliberate cultivation of attitude is also helpful in achieving such an objective.

20
Development of Scientific Attitudes

R. Will Burnett

If a hundred science teacher were selected at random and were asked to list their five chief teaching objectives, it is almost certain that every one would include the development of scientific attitudes and critical thinking on their list.

How well has conventional science teaching contributed toward the realization of these excellent objectives ?

Superstitions are evidences of unscientific attitudes. The person who believes that broken mirrors, black cats, and the number 13 bring bad luck possesses magical and uncritical beliefs. The degree to which science instruction has reduced superstitions is therefore one measure of its effectiveness in developing scientific attitudes. In a study of this area, Zapf[1] found that superstitious beliefs decreased slightly but significantly as the length of time students studied science increased.

On the other hand, Zapf[2] found, in an earlier study, that the mere teaching of science did not reduce superstitious beliefs and that only instruction dealing with specific superstitious appeared to be effective in reducing those superstitions.

There is considerable evidence that superstitious beliefs can be greatly reduced through instruction designed precisely for this purpose. But there is little evidence that science teaching, as such, will result in a mind-set antipathetic to superstitions. Yet, conventional science teaching, organized on the basis of an internally logical discipline, rarely devoted much time or attention to such beliefs.

That young people do go through our schools and science courses retaining many superstitious beliefs has been amply demonstrated. Caldwell and Lundeen[3] found that high school seniors believed in slightly more than 20 per cent of a list of superstitions and that they were apparently affected by about 22 per cent of the superstitious ideas with which they were familiar.

Ter Keurst[4] gave a check list of superstitious beliefs to over five hundred students in grades seven, eight, and nine. His check list contained ninety two beliefs, that had been judged to be of high significance by seven specialist in psychology. As a result of his study. Ter Keurst concluded that belief in superstitions does not decline with advance in grade level (despite the fact that the majority of his respondents were presumably taking or had taken science courses).

If one of the functions of science teaching is to help young people to examine their beliefs critically, then such studies as the foregoing must receive serious attention from science teachers, for they appear to present a failure of conventional science teaching.

Superstitious beliefs represent a tendency on the part of the believer to ignore the question of causation or to accept beliefs without demanding reasonable evidence of their validity. Other aspects of critical thinking and scientific attitudes have also been tested. Okes[5] examined explanations of natural phenomena from thirty five college faculty members, all of whom held at least a master's degree, but none of whom had majored in science. All these trained and intelligent teachers had, of course, some background of instruction in natural science. Yet, Oakes found that they showed little evidence of a consciously reflective or logical procedure in analyzing natural phenomena.

Alpern[6] developed tests through which he attempted to determine high school students ability to test hypotheses. He found that there was no significant relation between the science courses a student had taken and his ability to select sound procedure to test hypotheses. The degree to which Alpern's instruments were valid and his sample representative is the degree to which his study offers evidence that students have not developed this skill as a result of their science instruction. Unfortunately, there are no comparable studies reported in the literature.

A most revealing study by Lurie,[7] disclosed that the products of our schools are surprisingly liable to decide an issue, not on its merits, but on the basis of the prestige of proponents or opponents. Other studies comparable to Lurie's provide similar data. Such studies do not reflect the conventional science program alone, of course, but they do suggest that there may have been a considerable amount of formal or informal learning of an authoritarian nature. Common observation will disclose that many science teachers, along with teachers in other field, teach students to accept their statements without asking for evidence or support. Thus is laid a foundation for the student to accept propaganda uncritically.

A study by Davis of high school students and science teachers in Wisconsin disclosed that, although these students had a fairly clear concept of cause and effect relations and were not superstitious, they did not seem to be able to recognize the adequacy of supported causes in producing given results. Davis stated,[8]

> Many teachers tend to propagandize their material when there is no specific evidence for the statements they make, and teachers do not consciously attempt to develop the characteristics of a scientific attitude. If pupils have acquired these characteristics, it has come about by some process of thinking or experiences outside of the science classroom.

The conclusion that any increased power in critical thought processes or in scientific attitudes develops independently of, or possibly despite, science instruction was tested by Downing.[9] He administered a test of scientific thinking to approximately twenty-five hundred students in grades eight through twelve. He found, secondly, *that those students who had not studied science* secured higher average scores on the test of scientific thinking than did those who had studied science, juniors and seniors who had not studied science and who achieved average scores of juniors and seniors who had studied science from two to four years. It was Downing's conclusion that his study gave no evidence that science subjects, as they are conventionally taught, resulted in higher powers of scientific thinking. It is even possible to learnings in which the teacher and the textbook were assumed to be correct without supporting evidence might have developed an authoritarian viewpoint in the students and thus lessened of critical thinking.

The validity and reliability of the instruments and techniques that have been developed and employed in assaying scientific attitudes and abilities can, and should, be questioned. The degree to which they are valid and reliable is the degree to which studies such as the foregoing have demonstrated that conventional science teaching has not resulted in significant gains in scientific thinking on the part of the students. Whatever the tests and techniques measure, there is considerable evidence that newer practices in science teaching which emphasize the development of scientific thinking, produce students who are superior to those taught in conventional classes, as measured on these tests and by these techniques.

References :

1. Rosalind M. Zapf. "Relationship between Beliefs in Superstitions and Other Factors," *Journal of Educational Research*, 38 : 561–579 (April), 1945.
2. Rosalind M. Zapf, "Superstitious Beliefs," *School Science and Mathematics*, 39 : 54–62, 1939.
3. O. W. Caldwell and Gerhard E. Lundeen, "Students' Attitudes Regarding Unfounded Beliefs," *Science Education*, 15: 216–266 (May). 1931.
4. Arthur J. Ter Keurst, "The Acceptance of Superstitious Beliefs among Secondary School Pupils," *Journal of Education Research*, 32:673–685 (May), 1939.
5. Mervin E. Oakes, "Explanations of Natural Phenomena by Adults," *Science Education*, 29: 137–142 (April, May), 1945; 29:190–201 (October), 1945.
6. Morris L. Alpern, "The Ability to Test Hypotheses," *Science Education*, 30: 220–229 (October), 1946.
7. W.A. Lurie, "The Measurement of Prestige and Prestige Suggestability," *Journal of Social Psychology*, 9:219–225, 1938.
8. Ira C. Davis, "Measurement of Scientific Attitudes," *Science Education*, 19: 117–122, 1935.
9. Elliot R. Downing, "Some Results of a Test on Scientific Thinking," *Science Education*, 20:121 (October), 1936.

21

The Development of Scientific Attitudes

Richard E. Haney

In the Last few years considerable emphasis has been placed on the teaching of inquiry skills as well as on useful knowledge and manipulative skills. Some mention has been made of scientific attitudes, to be sure, but these still remain inconsistently defined in the literature and obscure in teaching in the literature and obscure in teaching plans. Daily lessons tend to center around some conceptual theme, a major principle, or some other form of cognitive learning outcome while affective learnings at best are considered peripheral to this central idea.

The habits of thought associated with scientific thinking deserve more careful consideration. Problem-solving skills are essentially amoral. Knowledge and intellectual prowess divorced from the controlling prowess divorced from the controlling influence of desirable attitudes towards man and nature contribute to the phenomenon which Robert Cohen termed the "frustration of humane living inherent in science of the twentieth century." (2) Science supposedly molds the character of its practitioners. To be scientific means that one has such attitudes as curiosity, rationality, suspended judgment, open-mindedness, critical-mindedness, objectivity, honest and humility.

Science lessons present many opportunities for teachers to help pupils develop these attitudes, which also have value outside the classroom and in other areas of human experience. Let us then consider the nature of attitudes, examine these eight attitudes and the overt behavior governed by them, and suggest appropriate learning experiences.

One of the most frequently quoted definitions of attitudes is the statement by Allport in which an attitude is described as a "mental and neural state of readiness, organized through experience, exerting a directive or dynamic influence upon the individual's response to all objects and situations with which it is related." (1) Sells and Trites points out that such mental and neural states cannot be observed directly in students. "An attitude is a psychological construct, or latent variable, inferred from observable responses to stimuli, which is assumed to mediate consistency and co-variation among these responses." (6) Attitudes regulate behavior that is directed toward or away from some object or situation or group of objects or situations. Attitudes have emotional content and vary in intensity and generality according to the range of objects or situations over which they apply. For the most part, attitudes are learned and are difficult to distinguish from such affective attributes of personality as interests, appreciations, likes, dislikes, opinions, values, ideals and character traits.

Guides to Scientific Behavior

A recent attempt to analyze the process through which attitudes are acquired appears in the *Taxonomy of Educational Objectives, handbook II: Affective Domain*. Attitudes are said to emerge first at the level of "willingness to respond" and become increasingly internalized in the learner through the stages of "satisfaction in response," "acceptance of a value," "preference for a value," commitment," and "conceptualization of a value. " At this last stage the learner is able to "see how the value relates to those that he already holds or to new ones that he is coming to hold."(4).

The first attitude to be considered is *curiosity*. This is the desire for understanding on the part of the student when confronted by a novel situation which he cannot explain in terms of his existing knowledge. A curious person asks questions, reads to find information, and readily initiates and carries out investigation. Curiosity is a stimulus to inquiry, and it is a desirable to inquiry, and it is a desirable outcome of instruction as well. Each discovery raises new questions and suggests new undertakings. Pupils should leave science courses with greater curiosity than they had at the outset. But, who are the most curious? Usually they are the younger children. Somehow our pupils manage to lose the spirit of inquiry with advancing age. Each teacher must ask himself how he can teach for heightened curiosity. Curiosity is learned. It can be learned or represented in the classroom. Problematic situations in which answers and explanations are not immediately available help to stimulate curiosity. The solutions of problems should raise new problems.

While curiosity stimulates inquiry, the attitude of *rationality* guides the scientist's behavior throughout his investigation. This is the habit of looking for natural causes for natural events. The rational person is not superstitious. The pre-scientific period in our history was marked by numerous examples of mythological explanations. This traditions ,still abounds in our folklore and in the everyday thinking of many persons. To help them develop the attitude of rationality, pupils can be confronted with situations in which careful reasoning proves superior to explanations of a superstitious nature.

Willingness to suspend judgment is another attribute of personality fundamental to scientific behavior. Persons with this attitude accumulate sufficient evidence before making judgment or drawing conclusions. They recognize the tentative nature of hypotheses and the revisionary character of our knowledge. To learn the attitude of suspended judgment, our students should be confronted with situations in which this behavior is rewarded or in some way leads to success while formation of conclusions without evidence leads to failure. Pupils should examine explicitly the consequences of jumping to conclusions.

Science teachers ought to examine closely the common practice of asking students to formulate a conclusion at the end of every five-minute demonstration or forty-minute experiment. These activities concern but a limited sample of all the phenomena governed by the principle under consideration. At the end of these experiences students should have the opportunity to choose among formulating a generalization with various qualifications, stating that they have only learned something about the particular operation at hand, or stating that they could make no sense of the data.

Acceptance of new ideas

Open-mindedness is closely akin to suspended judgment. To comprehend science as the human enterprise that it is, our future citizens must learn from experience that our ideas of what is true may change. They must be able to revise their opinions or conclusions in the light of new evidence. Experiences that foster open-mindedness include those in which pupils are confronted with the need to revise to belief as the result of having acquired new information on the subject.

The *willingness* to consider novel hypotheses and explanations and to attempt unorthodox procedures is a form of open-mindedness toward creative ideas which amounts to the "no holds barred" attitude of the scientist. The scientific method is not simply the application of routine of the scientist. The scientific method is not simply the application of routine and predetermined procedures to new problems. The study of new areas of knowledge often requires the invention of new methods of inquiry. Popular conceptions and explanations may fail to fit new bit of evidence. The history of science contains the stories of men who broke with traditions and saw nature in new light. To foster this creative spirit in the classroom, teachers can provide experiences in which pupils have the opportunity to design their own investigations and invent and evaluate their own explanations for natural phenomena.

New ideas are not accepted in science simply because they are new or different. To be scientific also means to be *critically minded*. A person with this attitude looks for evidence and arguments that support other persons' assertions. He challenges authority with the questions "How do you know?" and "Why do we believe?" He is concerned about the sources of his knowledge. One of the greatest temptations confronting the science teacher is that of giving direct answers to children's questions and of offering glib explanations. Teachers need to be careful of answers that include the word "because." Most explanations are not as simple as they might possibly appear at first.

How must teachers behave if their students are to learn the attitude of critical-mindedness? How often do they encourage their students to ask in class "How do you know?" To foster the learning of this attitude, teachers should provide evidence to support the generalizations in the lessons. Pupils should be taught to look for arguments and evidence supporting important propositions, and they should be taught to provide these in their own communications. The reading of historical and biographical accounts of investigations are also valuable experience from which pupils can learn of the sources of our current knowledge.

The scientist must also be *objective* in gathering and interpreting his data and intellectually *honest* in communicating his findings. To learn the attitude of objectivity, students may be confronted by situations in which the temptation to permit personal feelings to interfere with the recording of an observation or the interpretation of data must be successfully resisted in order to achieve a correct or accurate solution of problem. Complete objectivity is difficult to achieve because an observer's perceptions are governed by this previous experience and his expectations.

Intellectual honesty, on the other hand, is concerned with the conscious act of truthfully reporting observations. Teachers have to ask themselves how they reward honesty in their classrooms. In the laboratory, for instance, do the pupils know the *right* answers to report regardless of their actual sense data? The value of open-ended experiences for instructional purposes is that they are more like those of the scientist at the frontier of knowledge where the answers are not yet known. Science could not be the cumulative enterprise that it is if it were not for the objectivity and honesty of its practitioners.

Personal View of World

The foregoing attitudes directly govern the intellectual behavior of scientists and science students. To be "scientific" means to have these personality traits. In our classrooms, however, children learn more than the content and processes of science. They incorporate these bits of knowledge and skills along with those gained in other subjects and extracurricular experiences into their places in it. Each students gradually builds his own philosophy of life.

Humility is a desirable ingredient of the mature personality. It can be learned, at least in part, as a result of science instruction. Science can teach children to recognize their own limitations as well as the limitations of science itself. This is the attitude that underlies the conservation movement. It is the humble person who uses natural resources wisely, for the common good, even though he might have to forego immediate gains that could accrue from their exploitation.

Relationship to Nature

On the other hand, man's relationship to nature is more than a matter of "wise use." He shares this world with other beings whose "rights" deserve to be recognized. The history of science and also of religion is a story man's struggle with his own egocentrism. The message that Rachael Carson gave us in *Silent Spring* relates a current episode in that story. Albert Scheweitzer has expressed the attitude of humility in terms of "reverence for life" which identifies the moral principle that the good consists in the preservation, enhancement, and exhalation of life, and that destruction, injury, and retardation of life are evil. The word presents a spectacle of "will-to-live" contending against itself. One organism asserts itself at the expense of another. Man can only preserve his own life at the cost of taking lives. One who holds to the ethic of reverence for life injures or destroys only out of necessity. Never does such a person kill other beings from thoughtlessness. (5) There is a curious similarity in the messages of Rachael Carson, the scientists, and Albert Schweitzer, the theologian.

What sort of humility or reverence for life is taught in our biology classes? The trend, at present, is to increase the amount of experimentation with living materials in order to make the work of the student more like that of the scientist. But, perhaps, we have in "reverence for life" a limitation of discovery-method of teaching. What attitude do the students learn if animals are dissected merely to show that the chart on the wall or the plastic model on the demonstration table is correct? What concern for other lives is taught if an anesthetized rat is cut open so that the students can experiment with the heart of dying animal? Such activities are likely to reaffirm the attitude that of creation belongs to man to be plucked, manipulated, harvested, or controlled at his will for purposes *he* considers essential. To teach the attitude of reverence for life, it may be that various experiences will have to be employed to great extent, even thought to do so would be to compromise the principle of making learning experiences as much like those of the scientist as possible.

To Foster Attitude Building

The attitudes which have been explored are attributes of intellectually and emotionally mature individuals, persons who not only behave outwardly in desirable ways but who understand why they act as they do. If these and other attitudes are to be fostered, they must be planned for and not simply accepted as concomitants to cognitive outcomes. Klausmeier suggests eight steps that teachers can take to facilitate the learning of attitudes. (3) These may be interpreted in terms of the problems of science teaching in the following manner:

1. The attitude to be taught must be identified. Example of attitude related to science have been identified in this article.
2. The meaning of the vocabulary used to describe altitudes or the behaviors related to them must be clarified for the learner.
3. Informative experience about the attitude "object" should be provided. In the case of scientific attitudes these "objects" are usually the various situations that occur in the problem-solving process. Typical of these are(a) the sensing of the problem in a perplexing situation. (b) clarifying and the defining the problem, (c) formulating of hypotheses, (d) reasoning out the consequences of the hypotheses and the designing of investigations, (e) gathering of data, (f) treating and interpreting of data, (g) generalizing or drawing conclusions, and (h) communicating the results of the investigation to others. Students need to be structured in the performance of each of these steps and in their relationships to the various attitudes that characterize the scientifically minded person. It is hoped, of course, that pupils will exhibit these attitudes in appropriate situations outside the classroom. To help them generalize these attitude, teachers can point out the general nature of the attitude object by showing similarities between scientific problem-solving procedures and the treatment of problematic situations in daily affairs.
4. Desirable identifying figures for the learner should be provided. These models, whether they be teachers, parents, peers, or historical figures, provide the learner with ready made behaviour which he can use as his first attempts at the desired behavior.

5. Pleasant emotional experiences should accompany the learning of the attitudes. Pupils need freedom to attempt their own patterns of exploration and sufficient time to pursue an investigation to the point where they experience the satisfaction that accompanies inquiry and discovery.

6. Appropriate contexts for practice and confirmation should be arranged. Learning experiences must be selected of the basis of knowledge, skills, and attitudes to be learned. At times the central theme of a lesson might have to be a particular attitude with other learnings playing secondary roles.

7. Group techniques should be used to facilitate understanding and acceptance. The varied activities possible in well-equipped science rooms permit students to learn as individuals on some occasions and as members of groups of varying sizes on others. Group decision making that occurs in the planning and carrying out of investigations and the evaluation of results permits a sharing of emotional commitment which can enhance the learning of an attitude.

8. Deliberate cultivation of the desired attitude should also be encouraged. Pupils need to be aware of the behaviours that accompany an attitude and to practice them. Sometimes this requires the difficult task of breaking old habits or of improving poorly learned ones. The teacher must be able to provide guidance for this learning.

There are implications in what has been said for the education of teachers as well as for the instruction of school-children. It has often been said that "you can't teach something you don't know." A corollary to this generalization might be this: "Pupils cannot learn attitudes that their teachers don't have." It may very well be that the first step in meetings this challenge to science education will consist of an inward look upon our own knowledge and value systems. Science teachers have responsibility. It is to them that the public turns for an understanding of science, not just the facts of science, or the skills, but also for a perspective that relates science to all other areas of human experience.

References

1. Gordon Allport. "Attitudes." Chapter 17 in *Handbook of Social Psychology*, C. Murchison, Editor. Clark University Press, 1935. p.

806. Quoted in *Children's Thinking* by David Russell. Ginn and Company, Boston, Massachusetts. 1956. p. 170.

2. Robert S. Cohen. "Individuality and Common Purpose : The Philosophy of Science." *The Science Teacher*, 31 : 27–33. May 1964, p. 31.
3. Herbert Klausmeier. *Learning and Human Abilities: Educational Psychology*. Harper & Row, Publishers, New York. 1961. p. 267.
4. David Krathwohl et al. *Taxonomy of Educational Objectives, Handbook II : Affective Domain*. David McKay Company, Inc., New York. 1964. p. 36.
5. Albert Schweitzer. *The Philosophy of Civilization*, translated by C.T. Campion. The Macmillan Company, New York. 1959. pp. 307–329.
6. Saul Sells and David Trites. "Attitudes" in *Encyclopedia of Educational Research*, Chester Harris, Editor. The Macmillan Company, New York. 1960. p. 103.

22
Developing Scientific Attitude

Karuna Shankar Misra

"Development of scientific attitude" is an important objective of teaching science. Ours is a multi lingual and multi-cultural country. Scientific society is to be developed in order to help further citizens to adjust themselves and live without succumbing to the dangers of differing opinions or loyalties to different groups. Teaching of science should enable students to develop a personal philosophy of life based on truth, understanding and logic. Superstition, wishful thinking, blind following and unsystematic functioning are to be bid good bye. Our future citizens should not only possess scientific literacy but they should also possess scientific attitudes. This means that they should be open-minded in their pursuits, have the spirit of scientific enquiry, be able to observe and think critically and precisely, be unbiased and objective in their observation, have intellectual honesty, avoid taking decisions when ample evidences are not available, believe in the verified facts and be willing to change their opinions. Such characteristics are the essential ingredients of scientific attitude. Kozlow & Nay (1976) have talked of three components of scientific attitudes - (1) cognitive : manner in which attitudes manifest themselves in the professional behaviour of a scientist; (2) intent : tendency to show approval or disapproval of behaviours which define an attitude; and (3) action : extent to which students actually demonstrate in science classroom the behviours which define scientific attitude. Science teachers of many of our schools seldom pay any attention to these three aspects of scientific attitudes. They can play a vital role in shaping young vibrating minds if they try to develop scientific attitudes among students. They can adopt or adapt the following activities -

1. Developing curiosity

Curiosity can be learned as well stifled in science classrooms. Problem setting situation, useful for problem solving, can help in stimulating students' curiosity. When students find it difficult to explain a phenomenon on the basis of their acquired knowledge, they become more eager to know more. Interesting displays, bulletin boards and living organisms can be used to depict certain unusual things or events. This will create a dynamic environment for developing curiosity. Teachers may encourage students to ask questions and find answers to certain questions. They may try to discuss students' findings/answers in order to contribute to the origin of new questions. This will help in maintaining high levels of curiosity throughout the years.

2. Developing rationality

Teachers can develop student's rationality by asking them to give logical or scientific explanations or common superstitions. They can devise simple experiments to help students reject the superstitions. Video films, audio recordings or film strips from TV serials may be used for presenting situations in which persons exhibit their superstitions. They should be followed by brief sessions in which students given their reactions. Again the result of behaving in a superstitious way should be presented through video film, audio recording or film strips. Thus, the students will be able to evaluate the outcomes of situations and see the superiority and utility of logical reasoning to superstitious explanation.

3. Developing objectivity

While learning science students collect various kinds of informations and make impersonal judgements to explain them. Teachers may ask students not let their judgements be influenced by their personal feelings and thoughts. They may create problem situations that provide opportunities for collecting informations and drawing inferences. Importance of impartial research should be emphasized. Students may also be asked to analyze their response. Students' sensitivity to temptations should also be increased so that they may learn to resist temptations in favour of objectivity.

4. Developing willingness to suspend judgement

Students should not be eager to make hurried generalizations on the basis of inadequate evidences. Their willingness to defer judgements till ample evidence are gathered is very crucial. Teachers should provide opportunities which demand decisions on the basis of varying amounts of informations. Students should be made to understand the fact that inferences drawn from inadequate informations are inaccurate. Teachers can motivate students to suspend judgements and try to collect more informations.

5. Developing critical mindedness

Students possessing critical mindedness give more weightage to evidence for supporting statements. He respects evidence. Teachers can ask students to devised certain experiments to supply evidence to support or contradict their views or statements. Many times text books contain erroneous statements and diagrams. They can be used for fostering students' critical mindedness. Teachers may also set personal example by asking "How do you know? or why do you think so?" They must try to avoid making statements like the following -

— "I do not have ample evidence to give a reliable answer. However, I thinking ...."

— "Two out of these evidence tend to indicate that........"

— "Most of the scholars believe that.........."

While teaching for concept formation or concept attainment, teachers may present many positive exemplars that clearly depict essential attributes of a concept. They may ask students to looked for inconsistencies in concept definitions and inferences. Students must be taught to challenge the validity of unsupported statements and consult numerous authorities when seeking informations. Critical students should not be branded as disobedient or aggressive. Students who do not protest wrong explanations, facts or inferences should not be rewarded. It is usually seen that a student who doubts a statement of teachers or puts questions to them, is usually silenced, ridiculed or scolded. This injurious practice is to be given up. Authoritarianism can be substituted by democratic teaching and students can be allowed to question to clarify their doubts, if any.

6. Developing Open-mindedness

Students can learn to demonstrate their willingness to change opinions. Teachers can provide opportunities in which students think over new hypotheses and surprising processes. Students should be encouraged to find unorthodox solutions to traditional problems. Students should be made to alter their hypotheses when necessary to accomodate empirical data. Conclusions drawn by students should be discussed in the class and each student should be encouraged to evaluate evidences that contradict their own hypotheses or the ideas that are presented by others.

7. Developing Honesty

While conducting experiments many students try to guess the problem solutions and they do not report correct observations. Students can learn to express their unwillingness to compromise with the truth. Teachers may devise open ended experiments or investigatory projects so that students may not know the correct and appropriate problem solutions. Such experiences can develop students' self- confidence and honest reporting of all evidences even when they contradict their hypotheses or inferences.

The foregoing discussions have revealed that teachers can play a vital role in the designing learning activities that promote development of scientific attitudes. Teachers can create a congenial atmosphere in the class. They may not only allow students the freedom to ask questions, evaluate ideas and evidences, plan and execute experiments, formulate and test hypotheses, observe and report observations honestly and objectively, suspend drawing conclusions when adequate evidence is not available and change their opinions when contradictory evidences are available. Teachers should encourage students to know about the scientific attitudes as reflected in the historical background of scientists and their inventions. Demonstration of such attitudes in behaviour can be expected as well as positively reinforced by the teacher. Students can be encouraged to express their scientist like behaviours. Only necessary control should be exercised so that learning may occur in a friction free cohesive, organised, democratic, comfortable, resourceful and safe environment. All efforts will bear no fruit if students notice and feel that their science teachers have an aversion for scientific attitude or they seldom behave as scientists.

References

Kozlow, M.J. and Nay, M.A. An approach to measuring scientific attitudes. *Science Education*. 1976, 60(2), 147-172.

23

Types of Learning Experiences in Science Helpful in Developing Attitudes

Eldwood D. Heiss
Ellsworth S. Obourn
Charles W. Hoffman

Several lists of attitudes seem to be important as outcomes of science instruction have been made by science educators. There is considerable agreement among these list. The following list of attitudes is proposed as suggestive but any other list would serve equally as well. The science program should foster and develop the attitudes which will modify the individuals behavior so that he:

1. Looks for the natural causes for things that happen:
   a) Does not believe in superstitions, such as charms of good or bad luck.
   b) Believes that there is no connection necessarily between two events just because they happen at the same time or one after the other.
2. Is open-minded toward work and opinion of others and information related to his problem:
   a) Believes that truth never changes, but his ideas of what is true may change as he gains better understanding of that truth.
   b) Revises his opinions and conclusions in the light of addi-

tional reliable evidence.

c) Listens to, observes, or reads evidence supporting ideas contrary to his personal opinions.

d) Accept no conclusion as final or ultimate.

3. Bases opinions and conclusions on adequate evidence:

a) Is slow to accept as facts anything not supported by convincing proof.

b) Bases his conclusions upon evidence obtained from a variety of dependable sources.

c) Searches for the most satisfactory explanation of observed phenomena that the evidence permits.

d) Sticks to the facts and refrains from exaggeration.

e) Does not permit his personal pride, bias, prejudice, or ambition to pervert the truth.

f) Does not make snap judgment or jump to conclusions.

4. Evaluates techniques and procedures used and information obtained:

a) Uses a planned procedure in solving his problems.

b) Seeks to use the various techniques and procedures which have proved valuable in obtaining evidence.

c) Seeks to adapt the various techniques and procedures to the problem at hand.

d) Personally considers the evidence and decides whether it relates to the problem.

e) Judges whether the evidence is sound, sensible, and complete enough to allow a conclusion to be drawn.

f) Selects the most recent, authoritative and accurate evidence relates to the problem.

5. Is curious concerning the things he observes:

a) Wants to know the "whys," "whats," and "hows" of observed phenomena.

(b) Is not satisfied with vague explanations of his questions.

24

Techniques for Developing Scientific Attitude

K. Yadav

One of the major aims of teaching life sciences is the development of *scientific attitude* in the pupil. Following are some of the various aspects included in the scientific attitude:

i) Making pupils open minded.

ii) Helping pupils make critical observations.

iii) Developing intellectual honesty among pupils.

iv) Developing curiosity among pupils.

v) Developing unbiased and impartial thinking.

vi) Developing reflective thinking.

NSSE (National Society of the Study of Education) has defined scientific attitudes as "open mindedness, a desire for accurate knowledge, confidence in procedures for seeking knowledge and the expectation that the solution of the problem will come through the use of verified knowledge."

The views regarding scientific attitude expressed at a work shop conducted by the National Council of Educational Research and Training (NCERT) at Chandigarh in 1971 can be summarised as follows. A pupil who has developed scientific attitude.

i) is clear and precise in his activities and makes clear and precise statements.

ii) always bases his judgements on verified facts and not on opinions.

iii) prefers to suspend his judgement if sufficient data is not available.

iv) is objective in his approach and behaviour.

v) is free from superstitions.

vi) is honest and truthful in recording and collecting scientific data.

vii) after finishing his work takes care to arrange the apparatus, equipments etc. at their proper places.

viii) shows a favourable reaction towards efforts of using science for human welfare.

TECHNIQUES FOR DEVELOPING SCIENTIFIC ATTITUDE

In the previous pages an effort was made to define the terms "*scientific attitude*". By developing scientific attitude in a person certain mind-sets are created in a particular direction. Such mind-sets may be developed either by direct teaching in schools or by out of school experiences gained by the pupil. Though out of school experience contribute to a large extent yet according to *Curtis* direct teaching does modify the attitude of young pupil.

Tyler also made some suggestion for planning learning experiences in order to inculcate scientific attitude in the pupil. These are summarised below:

i) The increase in the degree of consistency of the environment helps in developing and including scientific attitude in the pupil.

ii) The scientific attitude can be inculcated in a pupil by providing him more opportunities for making satisfying adjustments to attitude situations.

iii) The scientific attitude can also be developed in the pupil by providing him opportunity for the analysis of problem or situation so that a pupil may understand and then rest intel lectually in desirable attitude.

Role of Science Teacher

The major role can be played by the science teacher in developing scientific attitudes among his students and this he can do by manipulating various situations that infuse among the pupils certain characteristics of scientific attitudes. He can also help in developing a scientific attitude among his students if he possesses and practices various elements of these attitudes. The practical examples given by the teacher leaves an indelible mark on the personally of his students.

Teacher can use one or more of the ways for developing scientific attitude among his pupils.

i) ***Making use of Planned Exercises:*** A large number of exercises for developing of certain scientific attitude are reported by various journals and magazines. Teachers can frequently use such exercises for developing certain scientific attitudes among the pupils. He can also make use of cuttings from newspapers and science magazines and can display such materials on bulletin board so that is used again and again for direct teaching.

   Exercises which are always included in good text books can also be used by the teacher for developing scientific attitude among his pupils.

ii) ***Wide Reading*** : On the basis of study conducted by him, Curtis reported, that those pupil who engage themselves in wide reading in science, develop scientific attitude more than those who study only one textbook. Thus a teacher should encourage his students to read library books and supplementary books on science. For this it is essential that each school at least has a science corner in its library. The teacher himself must be in habit of making proper use of science library so that the students get encouragement for use of science library. The teacher himself be familiar with the latest new titles in his subject and be willing to share his joys of new readings with his pupils. He should refer some suitable books to his students.

   Writing about teachers, Dr. Rabinder Nath Tagore has observed, " A teacher can never truly teach unless he is still learning himself. A lamp can over light another unless it continues to burn its own flame. The teacher who has come to the end of his subject, who has no living traffic with his knowledge, but merely repeats his lessons to his students, can only load their minds. He cannot quicken them".

iii) ***Proper use of Practicals Period***: A student of science gets many an opportunities learning scientific attitudes during his practical periods. It is for the teacher to properly use such opportunities for developing scientific attitudes amongst his pupils. Teacher should take extra care to state the problem of the experiment and should present hypotheses on solution. He should practice the proper method of testing the hypothesis. He should actively participate in discussion and interpretation of results after the experiment. He must inculcate in his students the habit to postpone judgment in the absence of sufficient evidence to support a hypothesis.

iv) ***Personal Example of the Teacher***: Personal example of the teacher is perhaps the single greatest force that is helpful in inculcating the scientific attitudes amongst his pupils. Psychologists have found a great tendency amongst the students to copy their teacher". It is therefore essential that science teacher is free from bias and prejudices while dealing his pupils. He should have an open mind and be critical in thought and action in his everyday dealing. He should be totally free from superstitions and unfounded beliefs and should be objective and impartial in his approach to his everyday problems. He should be truthful and should have faith in cause and effect relationship.

v) ***Study of Superstitions***: There are different types of superstitions that still prevails in Indian society. Simply talking of these superstitions and calling them bad and out of date, will not leave a lasting impression on the minds of the pupils. It will be more useful if the teacher can encourage at least a few of his students to carry out practicals on some popular superstitions such as that the presence of a broken mirror in any home leads to disharmony in that home or that if a cat crosses your way when you are going out for some work, then your work will not be done on that day etc. etc.

Such beliefs can easily be discarded by a student if he keeps a broken mirror at his home and finds to his satisfaction that it has not created any type of disharmony in his home. Similarly other superstitions and misbeliefs can be tested and easily discarded by a student of science. Various researches carried out in the field have drawn the same conclusion i.e. by practical survey and study of such common beliefs, students have developed permanent mind-sets or attitudes towards such superstitions.

iv) ***Co-curriculum Activities in Science***: Various co-co-curricular activities such as organising science club, hobbies club, science society, photographic club, organising scientific tours and excursions etc. can be taken up by science teacher. Such activities should be properly organised by science teacher under his direct supervision but students be given enough freedom to plan their activities. It will help inculcate in students some desirable scientific attitudes. Co-curricular activities may include making of scientific charts and models, making of improvised science apparatus etc.

vii) ***Atmosphere of the Class :*** A proper atmosphere in the class room provided a desirable atmosphere for inculcating of certain scientific attitudes in the pupils. By proper class atmosphere were mean that the room is properly arranged and suitably decorated in such a manner that it provides an incentive to the pupil to inculcate the habit of cleanliness and orderliness. In addition to such a congenial physical atmosphere of the class room, the teacher's behaviour also contributes to the development of proper class room atmosphere. For inculcating the scientific attitudes amongst his pupils teachers should encourage them in their various activities. He should also take care to see that his lesson contain are such as to encourage the students to ask a large number of intelligent questions. He should feel pleasure in answering and explaining such questions and must not snub his pupils for asking so many questions.

25

Acquisition of Certain Attitudes or Mind-sets

Eldwood D. Heiss
Ellsworth S. Obourn
Charles W. Hoffman

Science eductors have long recognized that scientific attitudes are among the most important outcomes which should result from science teaching. Although some educators have considered scientific attitudes by products or concomitant forms of learning there has been a persistent growing tendency to view these attitudes as equal to or superior to the knowledge objective of science instruction. Science teachers are becoming aware that if scientific attitudes are to develop from the study of science, they must be taught for directly and systematically in the same manner as we try to develop a mastery of the principles of sciences.

Science teacher should know and understand clearly what the scientific attitudes are. A person who is scientific:

1. Is curious about his environment.
2. Believes that every effect has a natural cause.
3. Is open-minded.
4. Is critical -minded.
5. Is determined not to believe in superstitions.
6. Is unwilling to accept as facts any statements not supported by convincing proof.

7. Is willing to change his beliefs upon presentation of new evidence.
8. Respects another's point of view.
9. Maintains such ideals as honesty, patience, persistence, fairness and thoroughness.

Success in developing scientific attitudes depends ultimately upon the teacher. And it is import that the science teacher keep always in mind that children form attitudes more from example than from abstract precepts. The teacher's intellectual honesty, willingness to admit error, listening to others' ideas, and dealing in an unbiased way with facts make a favorable and lasting impression upon pupils. The science teacher should live and act the part in order that warm conviction will accompany his teaching. The cultivation of the right attitudes requires both emotional and intellectual appeal.

What can a science teacher do to cultivate scientific attitudes ?

The following procedures are suggested :

1. Preserve democratic procedures in the classroom. If a desirable rapport is to be maintained between teacher and pupil there must be freedom of expression and a free movement of emotional influence.
2. Suggest project which give the pupils experience in problem solving. Suggest problems that require the collecting of evidence in advance of forming a conclusions.
3. Stress frequently the need for adequate data before arriving at a conclusion and that conclusions based on insufficient data should be accepted tentatively.
4. Pupils often have attitudes which are based upon error and misconception. Popular ignorance such as beliefs in superstitions, diet fads, astrology, patent medicines, self-doctoring and the like should be dealt with when the pupils reveal them. Naive attitudes, as pupils reveal them, may be selected for class discussion. For example, many children believe that handling a toad will cause warts on a person's hand. Questions like these may be raised by the teacher: What evidence is there that this belief is so? What kind of an experiment could be done to prove or disprove this belief? The goal of training is to develop an

open mind and a critical point of view. Try to create the attitude that statements are not necessarily true because they appear in print or because they are uttered by some older person in the community. Misconceptions can be dealt with most effectively by creating attitudes that will serve as checks on accepting half-truths and superstitions.

Science teachers must be patient. The difficulty in changing naive attitudes is related to their intensity. The more highly emotionalized a pupil's attitude is, the greater the difficulty in changing it. Children's superstitious beliefs will diminish with age as well as with educational attainment.

26

The Role of Science Teacher in Developing Scientific Attitude

Shaheen Usmani

Science teaching is an integral part of the school curriculum. In fact, the new era calls for a scientific orientation of all the subjects. A critical approach encompassing all the intrinsic values are being developed. Subjects which have been hitherto theorized are being reshaped in accordance with objectivity. Thus, science teaching becomes all the more important.

In order to make it meaningful, it has to be transacted earnestly. Nothing can be more harmful than science being taught badly. It will not only lead to misconceptions but will give negative contribution. The process of transaction has two important aspects-the curriculum and the teacher. However good the curriculum might be, it will undoubtedly fail to generate proper learning if the teacher is incapable of teaching. The success of teaching learning strategies adopted depend on the teachers competencies.

Satisfaction and Annoyance effect Attitudes

The teacher plays an important role in developing the right attitude towards science. The teachers in schools make many children practise things with annoyance. They spend much of their time in forcing their students to go through the form of learning right things but almost no time is provided for the right attitudes on the part of the students. If the children are practising right things with annoyance, they are unlearning those right things, not learning them. The attitudes and feelings of the students in their practice are to negligible, they are of the utmost importance in all their learning.

To develop scientific attitude it is important that science teaching itself provides satisfaction. The ultimate success of seeking truth in highly contending. The teacher should know that children learn to follow the ways that succeed and give satisfaction. They learn not to follow the ways that fail and give annoyance. Teachers need to provide their students with large opportunities for practising good attendant learning with success.

Significance of Science Teacher

We no longer accept the fact that the teacher is one who "teaches". Today a true teacher is one who learns and inspires. Effective learning is possible when the teacher possess certain competencies to make learning simple and interesting.

For the task of spreading knowledge and building up habits of thought and action, it is important that the science teacher is well qualified and aware of the various modes of transporting knowledge. An efficient and resourceful science teachers can carry out his work quite efficiently even with inadequate science facilities.

The role of teacher and his importance has been highlighted by various eminent educationists. Some of them are given below:

> "*Teachers should be filled with a modernist—attitude, with a progressive, with a forward looking direction. Unless they themselves have it, they cannot make their students forward looking*".
>
> **—Sarvapalli Radhakrishnan**

> "*A teacher has to help in the transmitting of higher values to his pupils, through his personality and through the goods of culture which are his instruments. A teacher has to help the bud into full bloom and not to make paper flowers to satisfy his whim. The growth of a morally autonomous personality is the aim and end of his endeavour*"
>
> **—Jawahar Lal Nehru**

"A teacher can never truly teach unless he is still learning himself. A lamp can never light another lamp unless it continues to burn its own flame. The teacher who has come to the end of his subject, who has no living traffic with his knowledge, but merely repeats his lessons to his students can only load their minds. He cannot quicken them. Truth not only must inform but also inspire. If the inspiration dies out and the information only accumulates then truth loses its infinity. The greater part of our learning in the schools has been a waste because, for most of our teachers, their subjects are like dead specimens of once living things with which they have learned acquaintance, but no communication of life and love".

—Rabindra Nath Tagore

Scientific attitude and the Science Teacher

Scientific attitudes is a tendency to seek truth, think logically and there upon act reasonably. To develop scientific attitude among students, the teacher has to be eager and full of enthusiasm.

The National Policy of Education (NPE) 1986 has also emphasized the generation of scientific attitude among students through its curriculum and teacher. The following statements on science education in Part-VIII of the document dealing with Re-orienting the content and Process of Education have been outlined:

i) Science Education will be strengthened so as to develop in the child well-defined abilities and values such as the spirit of inquiry, creativity, objectivity, the courage to question, and an aesthetic sensibility.

ii) Science Education Programme will be designed to enable the learner to acquire problem-solving and decision making skills and to discover the relationship of science with health, agriculture, industry and other aspects of daily life. Every effort will be made to extent science education to the vast numbers who have remained outside the pace of formal education.

Good teachers are needed as much as possibly more. They are need to guide and encourage the children and arrange situations where the children will educate themselves. A teacher who is properly guiding his students can help to build interest in good things and approval of them. The newer methods acquire that teachers shall be able to see that the students think for themselves. In order to have children think for themselves, time and patience and necessary. They are often ready to guess, when they are given the chance, rather than to think. Practice in thinking is hard to provide for others. In order to do so, teachers themselves need to practise thinking.

In addition to the personnel qualities, a science teacher should possess the following qualities:

1) Basic academic qualifications.
2) Trained in the modern methods and techniques
3) Practical knowledge of child psychology.

Competencies

Certain competencies have been identified from time to time in order to achieve the objective of Science teaching. Scientific attitude can be inculcated among the students through the following teaching competencies:

1) Develop in the students the concept of integrated science so as to the help the students acquire sound scientific literacy and appreciate social ethical aspects of science.
2) State instructional objectives in terms of specific behavioral outcomes.
3) Plan suitable activities select appropriate resources and organize group activities.
4) Design and practice activity based on learner centered approaches of teaching.
5) Develop micro-teaching skills and skills of organizing out of school or extended curricular activities such as science clubs, science fairs, science exhibitions and field studies.

Science should be taught through the procedure of inquiry. The students will learn that science is not memory or magic but rather a disciplined form of human curiosity.

Activity Based Approach

This approach has to be systematic and well organized in order to be affective. Among the techniques of instruction which play an important role in the type of effective curriculum transaction involving activity based approach, the teacher has to be apt at

- Planning of activities
- Preparing the students for activities
- Conducting and supervising activities
- Conducting discussion
- Designing activities for evaluating learning outcomes.

i) Planning of Activities involves that the teacher should be equipped suitable to understand the general objectives of the course, the specific objectives of each unit, work out time schedules and material check list. Though a teacher can perform this task all by himself, it would be ideal to involve the students in this.

ii) Preparing the student for activity involves introduction of the subject, identification of the problem, evolving a strategy for solving the problem, getting the children and material organised.

iii) Group discussion involves pooling of various ideas from children, leading to a discussion, encouraging and stimulating creative thinking.

iv) Activities for evaluating learning outcomes my be organised under the following three broad areas:

   a) Operational/functional understanding of scientific principles/concepts.

   b) Competence to solve problems and

   c) Scientific attitudes, interest and appreciations.

Such activities may include; teacher's observations, situational problem solving, participation of the pupils, assignment of projects, short quizzes, paper and pencil test, and children's performance at work etc. The science teacher should use scientific knowledge in erasing false belief, prejudices and practices prevailing in the society. He should encourage decision making problem solving skills in daily life situations.

Teacher as organizer

The teacher task is to select activities and materials to be used during the exercise. He has to organize the whole process of experimentation.

Teacher as facilitator

The teacher should facilitate the teaching learning process. Introduction of the topic, questioning technique, student teacher relations all put together create a conducive classroom climate for developing scientific attitude.

Teacher as moderator

The teacher has to lead discussion and moderate the classroom activities.

Teacher as a guide

The teacher has to identify the slow learners and choose appropriate remedial strategies using peer group learning.

Few suggestions to science teachers

1. The Science teacher should be well aware of not only the subject matter but of the innovations and new techniques.
2. He should make maximum use of all the immediate resources like laboratory, manuals and reading materials available in school.
3. Finally his teaching should be activity based involving students.

References

1. Kilpatrick, W. H., *How we learn, Y.M.C.A.* 1933. Publishing House, 5, Russel Street, Calcutta.
2. Sharma, R.C., *Modern Science Teaching*, 1984, Dhanpat Rai & Sons, 1982, Nai Sarak, Delhi -110006.
3. NC.E.R.T. New Delhi. - Developing course outline for the teaching of Science in the Teacher Education Curriculum.

27
An Attempt to Measure Scientific Attitudes

Howard B. Baumel
&
J. Joel Berger

SCIENCE educators have long recognized that scientific attitudes are among the most important outcomes which should result from science teaching.[1] There is general agreement among investigators that a person who has a scientific point of view (1) looks for the natural causes of events: (2) is open-minded towards the work and opinion of others and towards information related to his problem; (3) bases opinions and conclusions on adequate evidence; (4) evaluates techniques and procedures used and informations obtained; and (5) is curious concerning the things he observes.[2]

Although some educators have recognized the scientific attitude as products or concomitant forms of learning there has been a growing tendency to view these attitude as equal to, or superior to, the knowledge objective of science instruction. Science teachers are becoming aware that if scientific attitudes are to develop from the study of science, they must be taught directly and systemically in the same manner as a mastery of the principles of science is developed.[3]

Thus an evaluation programme which attempts a complete appraisal of a student s growth in science has as one of its obligations the use of some technique to reveal growth in the scientific attitudes.

Much experimentation has been carried on in the field of measuring attitudes and opinions. Literally thousands of articles and not a few books have been written on the subject. Most of these have come from the sociologists and the social psychologists. Consequently, a large proportion of the literature deals with the measurement of attitudes on social questions. Science teachers are not unaware of the need, however, of some valid and reliable test of scientific attitudes.

Judging from their prevalence in the literature, three studies in the measurement of scientific attitudes seem to be most widely accepted by science educators. These studies may be briefly summarized below.

Ira C. Davic[4] devised a test of 66 items which had for its purpose to test for the "possession of the concept of cause and effect relationship and of the ability to distinguish between fact and theory" in high school pupils and teachers in Wisconsin, thereby determining whether they possessed these characteristics of the scientific attitude.

The tests constructed and evaluated by Victor H. Noll[5] contained items providing opportunities for exercise in the following "habits of thinking" (1) accuracy; (2) intellectual honesty; (3) open-mindedness; (4) suspended judgement; (5) looking for cause and effect relationship; and (6) criticalness, including self-criticism.

A.G. Hoff[46] constructed a test for, and attempted to measure the degree of, the scientific attitude achieved by high school senior according to their major study. The purpose was to determine if those who had studied science as a major subject would score higher on the scientific attitude test than those who had pursued a major study in other areas.

Other investigators, including Riener, Howard and Robertson, and Davis also have contributed to this field.

There have been few attempts in recent years to develop instruments measuring scientific attitudes which reflect the impact of science upon daily life. The tests widely known in the field are largely outdated in terms of pupil experience and scientific progress.

It has therefore seemed desirable to construct a new testing instrument which attempt to keep within the range of experience of the students, avoid the expected response type of question, and eliminate the need for a complicated response on the part of the student.

The following questions are representative of some of the items included in a test designed to measure scientific attitude in ninth year general science students:

Type I

To measure, student ability to distinguish between scientifically determined facts and commonly held beliefs. (To be answered true or false.)

a) Students who do well in their studies tend to be poor in athletics.

b) There is a relationship between number testing and the level of radioactivity in the atmosphere.

Type II

To test student ability in evaluating experimental situations in terms of evidence available for valid conclusions. (To be answered in terms of enough or not enough evidence.)

a) Group A of rats received a diet consisting of white flour, butter, dried beef, starch, salt and calcium carbonate. Group B of rats received the same diet plus yeast. After several weeks, of the rats of Groups A showed loss of weight, rumpled fur, and other signs of unhealthy animals. The yeast must contain something which is necessary for the good health of rats.

b) Brom thymol blue is a chemical indicator which takes a yellow color in the presence of carbon dioxide. A test tube containing a green plant, carbon dioxide, and brom thymol blue is placed in the sunlight. After several hours, the solution in the test tube is obtained to have changed to a blue color. This ....indicates that carbon dioxide is no longer present in the test tube.

Type III

To test student ability to formulate conclusions based upon sufficient evidence. (To be answered true or false.)

a) Pellagra is a disease caused by a lack of vitamin called niacin. A child came down with a disease diagnosed as pellagra. It can be concluded that the diet of this child did not consist of food which contained a satisfactory amount of niacin.

b) A green plant was kept out of the sunlight for two days. A leaf was taken from this plant and boiled in alcohol. Alcohol dissolves out the greet coloring matter. After five minutes the leaf was observed to have turned white. This was because the plant was kept out of the sunlight for so long.

Type IV

To measure student ability to suspected judgement in non-experimental situations. (To be answered true or false.)

a) Scientist E, who is very well-known for discoveries he made several years ago, performed a certain experiment. Scientist D, Who just graduated from college, performed the same experiment and obtained different results. The results of Scientist E should be accepted.

b) A boy had a irritation his hand. His doctor prescribed an ointment and in several days the irritation disappeared. A week later, this boy's younger brother developed an irritation which looked similar to the one the older brother had. The younger boy should use the ointment the doctor had prescribed for his older brother since it worked so well for him.

The test was administered to ninth year students at the end of the year's work in general science. The results revealed that : (a) students who scored high were not necessarily those with high grades in science; and (2) students who scored low were not necessarily, those with low grades in science.

From this study emerge a number of hypotheses which are subjected to further investigation:

1. Scientific attitudes may be acquired by students at all levels of ability.
2. The science teacher needs to evaluate not only the knowledge achievement of his students but also their growth in scientific attitudes.
3. The students with scientific attitudes will more effectively cope with problems in school and community.
4. Success in developing scientific attitude depends ultimately on

the teacher. The teacher through his actions must be able to convince the students that scientific attitudes are an integral part of his behavior. His intellectual honesty, willingness to admit error, listening to others' ideas, and dealing with facts in an unbiased way make a favorable and lasting impression upon pupils.

References

1. Sam S. Blane, "Review of the General Goals in Science Teaching." *Science Education*, 36:47 (February, 1952).
2. *Rethinking Science Education*. Fifty-nenth Yearbook of the National Society for the Study of Education, Chicago. Illinois: University of Chicago Press, 1960. p. 46.
3. Nathan S. Washton, *Science Teaching in the Secondary School*. New York : Harper & Brothers, 1961, p. 39.
4. Ira C. Davis, "The Measurement of Scientific Attitudes," *Science Education*, 19:117–122 (October 1935).
5. Victor H. Noll, "Measuring the Scientific Attitude," The *Journal of Abnormal and Social Psychology*, 30:145–154 (July–September 1935).
6. A. G. Hoff, "A Test of Scientific Attitude," *School Science and Mathematics*, 36:763–70 (1936).

28

The Measurement of Scientific Attitudes

Ira C. Davis

In speaking of the measurement of science attitudes, I cannot give you a complete picture of what we are trying to do a Wisconsin without giving the background of our whole program. The measurement of scientific attitudes is only one future.

A few years ago the Teacher Training Council in Wisconsin wrote a philosophy of education. This has received wide recognition and is acknowledged as one of the outstanding statements of the purposes education. As a follow-up to that philosophy of education a committee of science teachers was asked to interpret this philosophy in terms of science teaching. In this general philosophy "education is recognized as growth through problem solving so that the individual will act in such a way that he will make the greatest contribution to society and at the same time receive the greatest personal satisfaction." What specific contributions can science make in the accomplishment of this general program?

The Science committee was faced with the problem of deciding what method it was going to use in developing this philosophy. After many conferences and discussions, we finally decided we would go out to the science teachers in the State and get their reactions, opinions, and emotional responses, rather than digest available literature. No doubt this literature has had a great deal of influence on the thought reactions of these teachers, but we were more interested in what they really thought themselves, what things they were in favor of and what they opposed.

In order to put these plans into operation, the committee organized more than 350 teachers into fifty-five different groups. The whole program was and still is cooperative. The groups held meetings. At regular intervals, summaries of reports were sent to the group leaders. The committee members attended meetings with as many groups as possible, (actually 30 different groups.) After a year's discussion, the committee met for three weeks and wrote the philosophy of science teaching.

In this philosophy we recognize the purpose of science is to develop the ability in the individual to solve the problems that confront him. To do this, he will need (1) a scientific attitude, (2) a scientific method of procedure and (3) a fund of information which will make it unnecessary for him to repeat what scientists have done before him. We opposed ranking objectives because all of them are important, and they are so closely interrelated that it is exceedingly difficult to say where one starts and the other ends.

We also listed fourteen specific objectives without any attempt to rank them. Neither did we try to place them in separate categories, such as attitudes, methods, interests, appreciations, because, as far as we knew, there was no commonly accepted line of demarcation between such categories. The fourteen specific objectives are:

1. Command of factual information.
2. Familiarity with laws, principles and theories.
3. Ability to distinguish between fact and theory.
4. Concept of cause and effect relationship.
5. Ability to make observations.
6. Habit of basing judgment on fact.
7. Ability to formulate workable hypotheses.
8. Willingness to change opinion on the basis of new evidence.
9. Freedom from superstitions.
10. Appreciation of the contributions of science to our civilization.

11. Appreciation of natural beauty.
12. Appreciation of man's place in the Universe.
13. Appreciation of the possible future developments of science.
14. Possession of interest in science.

It is not to be assumed that the philosophy is final. We had to begin somewhere. It is our working plan for the time being, and it will be revised as rapidly as we discover evidence which will help us revise it. The question is: Is this philosophy possible of attainment in the science classrooms? Will it work? If it does not work, what is wrong with it?

In order to answer these questions it becomes necessary to discover what is being accomplished in the classrooms. To do this it is necessary to have measuring instruments which will measure what is being accomplished. So our testing program in time will test for all of the objectives we have mentioned, both general and specific. With the exception of the fact or knowledge tests, we must explore the field in the testing of attitudes, methods, appreciations and interests. We really have no beginning ground; nor are we convinced that the so-called fact or information tests have reached their final forms. So we are constructing measuring devices as rapidly as we can proceed and still feel some security in what we are trying to do.

No doubt you are beginning to wonder when I am going to say something about scientific attitudes. I thought that presenting our plan first would lead you to appreciate more fully the difficulties we have had to face. There is much confusions about scientific attitudes. When does a person possess a scientific attitude or attitudes and when doesn't he ? What are the characteristics of a person if he does have a scientific attitude? As I have stated before there was no satisfactory definition for a scientific attitude in the dictionary or anywhere else. Neither did we feel that a definition made by the committee would be generally accepted. We could see no reason why it should be. But we had to begin somewhere. We analyzed carefully the statements of Downing, Noll and Curtis because they were the most recent. Downing thinks a scientific attitude a an urge to do something. You either have this urge or you do not have it. As far as we could see they made but little distinction between emotional reactions and scientific attitude. Noll says scientific attitudes and scientists thinking are the same thing. Curtis' statement is something like this—While scientific attitudes and scientific methods are so necessity closely related and inseparable they are nevertheless distinctly different concepts. Many other statements were analyzed carefully in the same way with no more satisfactory results. However, we did succeed in getting a number of characteristics which might be considered as statement of a scientific attitude.

After much discussion we decided to send this list of characteristics, with a few more added which were contradictory to each other, to 250 well-trained, experienced teachers in all parts of the United States 100 of them being sent to Wisconsin teachers. We recognize the questionnaire method is not entirely satisfactory, but we should yield better returns than more guessing. We received 162 replies, 92 of which we kept for final statistical treatment. In many questionnaires of this nature the tendency is to agree with the intent of this persons making out the questionnaire. The teachers marked all of the statements as being characteristics of a scientific attitudes they were discarded. They had made no attempt to discriminate between the items. They agreed with the printed page.

We ranked the characteristics as selected by these experienced teachers, and finally accepted those which were chosen by at least 80 per cent of them. These are the characteristics which were finally selected:

1. Willingness to change opinion on the basis of evidence. (92%)
2. Search for the whole truth regardless of personal, religious or social prejudice. (89%).
3. Concept of cause and effect relationship. (86%)
4. Habit of basing judgment on fact. (85%)
5. Power or ability to distinguish between fact and theory. (82%).
6. Freedom from superstitious beliefs. (81%).

We can say then that an individual who as scientific attitude will (1) show a willingness to change his opinion on the basis of new evidence; (2) will search for the whole truth without prejudice; (3) will have a concept of cause and effect relationship; (4) will make a habit of basing judgment on fact; and (5) will have the ability to distinguish between fact and the theory. That is our definition of scientific attitude for the present.

The next task was to devise valid tests for determining the presence of each of these characteristics in an individual. There is the danger that when tests of attitudes are given the pupil will respond as he believes the teacher wants him to respond, rather than in a manner consistent with his own inclinations. It would be much easier to test for these characteristics orally if a teacher had time to do it, but under present conditions it could not be satisfactorily done in any large scale.

The first test we constructed was the cause and effect relationship test. Not only is it essential to know whether or not pupil realizes that for any effect there must have been a cause, but it must also be ascertained whether the pupil recognizes the adequacy of a supposed cause to produce the even result.

We have 65 items in our present cause and effect relationship test. We have pointed 66 pairs of occurrences, and we asks the pupils to judge each of the paired occurrences by checking it as:

A— if the first occurrence is practically the sole cause of the second.

B— if the first occurrence is one of a number of the important contributing causes of the second.

C— if the first occurrence contributes only slightly to the second.

D— if both occurrence are results of the same general cause or causes.

E— if the first occurrences bears no casual relationship to the second.

A few of the paired occurrences follow:

1. The sun shines on the earth; the earth is warm.
2. A boy often picked up toads; the boy had warts on his hands.
3. The light of lightning; the accompanying thunder.
4. The ignition switch of an auto is turned on; the motor starts running.

5. A rising column of air was cooled; a cloud formed.

The test was mimeographed and tried out with 295 pupils in six different high schools. The number of pupil's answers for each of these five divisions has been determined.

The committee felt it needed help in establishing the best answer, or acceptable answers, for each list of paired occurrences in the test. So we sent the test to 40 outstanding teachers of science and asked them to score the tests. We receive the answered tests from twenty-five of these teachers, fifteen from outside of Wisconsin and ten from Wisconsin. (More papers were returned later.) We have answers from 295 pupils and 25 well-trained, experienced teachers.

A fact theory test was also constructed. This contained 103 statements. It was given to the same 295 pupils and scored by the same 25 teachers. The pupils were asked to check each statement for one of the following answers.

A. Some are statements of well established facts which are always true.

B. Others may be statements of well established theories which are generally accepted.

C. Others may be statements of theories which are questioned by some (many) authorities.

D. Others may be statements of popular beliefs, which are not supported by evidence.

E. Column E is to be checked if you are not familiar with the statement.

Following are some of the statements used for testing:

1. A disease is a punishment for some particular moral wrong.
2. Air is composed of molecules.
3. The pressure in water varies with the depth.
4. Heating the molecules in air increases their speed.
5. A high forehead indicates high intelligence.

On the basis of these preliminary tests, some of which were given to teachers as well as to pupils, we feel we are justified in drawing some tentative conclusions;

1. High-school pupils in Wisconsin are not superstitious.
2. High-school pupils make almost as good records as the teachers.
3. Many of theories in science are being taught as facts by many of our best teachers. Teachers as well as pupils fails to distinguish clearly facts from theories.
4. Pupils seem to have a fairly clear concept of the cause and effect relationship, but they do not seem to be able to recognize the adequacy of a supposed cause to produce the given result.
5. Many teachers tend to propagandize their material when there is no scientific evidence for the statements they make.
6. Teachers do not consciously attempt to develop the characteristics of a scientific attitude. If pupils have acquired these characteristics, it has come about by some process of thinking or experiences outside of the science classroom.

As a result of the suggestions of many teachers and an analysis of the statistical results, the first forms of the tests were revised. We now have them in printed form.

In September we gave tests to approximately 1000 pupils in eight high schools in Wisconsin. These tests were given in each school to classes in general science, biology, chemistry (if taught), physics and a class in which none of the pupils were enrolled in a course in science. We are going to give the same tests to the same pupils at the end of the school year next June. As a result of these tests we may be able to discover:

1. How well pupils are able to distinguish between facts and theories and how much they improve during the year.
2. How well pupils understand cause and effect relationships and what improvement they make ?
3. How much each science subject contribution to the development of the different characteristics of a scientific attitude?

4. How much the number of courses in science contributes to the development of attitudes? Are two or three courses better than one?
5. What progress do pupils no enroll in science make as compared with the different science groups?
6. Later we hope to show what effect different methods of teaching will have on the development of the characteristics of scientific attitude.

We are now using the following method of treating the result statistically. I am not at all certain this is the correct method. We are assigning different values, for each item ranging from plus 2 to minus 2. If the pupil marks the question according to the accepted answer, he will be given a score of plus 2. If he marks an answer that could not possibly be the right answer he will be given a score of minus 2. If his answer if fairly reasonable, he may be given a mark of 0 or plus 1. These assigned values are based on the results made by the pupils in the preliminary tests and on the answers which teachers have given. We have also checked each statement in the tests for accuracy in textbooks and source books. It is possible then to get a score of plus 2, plus 1,0, minus 1, or minus 2, on any item. It is believed that we will get a more refined statistical treatment by using these assigned values than we would by simply checking the answers as wrong or right. We hope to know more about the use of these assigned values when we have completed the statistical treatment for the 2000 tests we gave in September.

We are laying the groundwork for the construction of the other tests. We are having difficulty in getting the exact type of test we think we should use. In testing for willingness to change opinion on the basis of evidence we must discover;

1. What a pupil's present opinion is.
2. How much evidence does an average pupil need to have his opinion changed?
3. What is the best way to present his evidence?
4. How will we know that he has really accepted the evidence presented? How can we overcome the notion which is so common, "You believe what you want to believe"?

5. How will we know when the pupil has really changed his opinion?

One type of question we are experimenting which now is the following. (A) A man had his car oiled in a garage and after driving a few miles discovered that the bearings in his car had burned out. Should the owner of the car attempt to recover damages from the garage owner? Many other situations may be stated which include additional evidence, such as, (B) On examination the crank case was found to be full of oil; (C) The crank case was empty; (D) The cap which holds the oil in the crank case was found in the garage; (E) There was a crack in the crank case.

If a series of changes like these were given it should be possible for us to discover what a pupil's opinion is, how well he weighs evidence, and how well he is able to distinguish between the type of evidence given in the different situations.

In measuring willingness to change opinions on the basis of evidence, it becomes necessary to discover if the pupil can judge evidence in any particular case, not evidence in general. This ability to judge evidence in any particular case will depend to a large extent on a knowledge of facts in that case. If we can assume a pupil has a general knowledge of the facts in any case, the, it should be possible to present experimental evidence as discovered in recent researches and discover how the pupils react. Much of the information in our textbooks is inaccurate. Some of it may be easily contradicted by recent experiments.

A few statements in many of our textbooks may be used for illustration:

1. Wood decay by oxidation, or there is oxidation during decay. (Recent experiment seem to prove there is no oxidation during the decay of wood. This decay is caused entirely by fungi or other plant diseases.)
2. Squirrels burry nuts so they will have a supply of food for the winter, and the squirrels remember where they bury the nuts. (The evidence for this is the opposite from the statement made in many of our textbooks.)

Again, many statements are made ascribing certain intellectual features to definite physical characteristics in individuals. Some for the statements follow: (1) "A high forehead indicates high intelligence"; (2) "A square jaw indicates a strong will", (3) "Red hair indicates a fiery temper."

Also, many statements in our health and physiology textbooks are inaccurate and have been made without much evidence to support them. In too many cases conclusions have been drawn on too few instances or on the personal opinion of some individual who could not be considered an authority under any consideration. In cases like these, we would give the statements in textbooks, ask for the student's opinion, then present the evidence, and ask them to interpret the evidence in terms of the original statements.

Our present methods of advertising provide a fertile field for the judgment of evidence. Let pupils read some advertisement, ask them to analyze the statements in the advertisement under the following divisions:

a. Statements of absolute-facts-outright.

b. Statement of such a nature that the advertiser wants you to think the statement is a fact, and which is worded in such a way that the facts are not stated but intimated. The statement actually is a misstatement of fact.

c. Statements which are facts but they have no connection with the advertisement. The statements are irrelevant as far as the advertisement is concerned.

d. Statements in which evidence of some experiment is presented which is intended to lead the reader to believe that the experimental evidence is the deciding factor in making the advertised product worth while. The experimental evidence presented may have no connection with the so-called values of product. For illustrations:

"It dissolves in two minutes." What if it does?

"It cannot harm the heart." What good does it do?

"Five cups will not harm you." What good does it go? Why five?

You see advertisers know that people have faith in the results of experiments. They are using the methods of the scientists to mislead people, rather than give a true statement of facts. I am not inferring that all of our advertising is dishonest but too much of it is of such a nature that it tends to give values to products that do not have the values claimed for them. We must be on guard then to give pupils training in detecting falsehoods (lack of evidence) in the factors of life which depend so much upon honesty. They must learn how to judge evidence. They must train themselves to form their opinion on the basis of evidence. But pupils do not get this training by just learning the facts in science. They must be taught how facts are obtained, what facts are relevant in a situation, and how fact my be used to form correct opinion.

Three of the characteristics of scientific attitude are similar and centre around the idea of coming to a proper conclusion from evidence, or facts, rather than holding a prejudiced opinion. "Will ingness to change opinion on the basis of evidence", "Search for the whole truth without prejudice," and "The habit of basing judgment on fact," can probably be tested with the same test. By definition, "Prejudice is an opinion or judgment formed beforehand" It is a preconceived judgment or opinion ""It is the learning toward one side of question from other considerations than those longing to it." "It is the unreasonable objection against anything." "It is an opinion or learning adverse to anything without just grounds or before sufficient knowledge has been obtained." If a pupil has the habit of basing judgments on facts, he will search for the whole truth without prejudice and will be willing to base his judgements and opinion on evidence and facts.

It is not to be assumed that by adding the scores on these three tests we will have a single index of a pupil's scientific attitude. They are rather tests of characteristics inherent in a scientific attitude. We cannot assume that the scores on the individual tests are positively correlated. Measurement of a scientific attitude can only be achieved by the careful determination of specific elements.

The ultimate aim is not testing but improved teaching. We must convince science teachers by clear-cut evidence that teaching science for the sake of knowledge alone is not enough. We have no quarrel with teachers who insist on teaching scientific knowledge. We doubt that we can make advances in teaching scientific methods and attitudes without a good background of knowledge. But we feel that teaching knowledge alone is not enough and that we are neglecting many opportunities to give pupils the real values of science instruction.

29

An Approach to Measuring Scientific Attitudes

M. James Kozlow
Marshall A. Nay

INTRODUCTION

Lists of objectives proposed for science education often include the development of interests, attitudes, values and appreciations. However, the literature on this topic [1,2] indicates that teachers tend to neglect these objectives when planning classroom activities.

In their discussion of summative and formative evaluation, Bloom and coauthors[1] discuss possible reasons for this avoidance of direct attention to affective objectives. They claim that there is a general feeling, among both the public and the teaching profession, that trying to develop selected attitudes and values in students in skin to indoctrination and brainwashing. However, Scriven maintains that teaching toward effective objectives should be approached in a manner similar to the teaching toward effective object should be approached in a manner similar to the teaching of science concepts. Selected values and attitudes should be presented by the teacher as the most defensible ones from given set of alternatives. Emphasis should be placed on developing an understanding of the arguments in support of those which are selected.

Another contributing factor to the neglect of affective objec tives is the inadequacy of available teaching methods and materials for the development of interests, attitudes and values. Finally, the dearth of suitable evaluation instruments of techniques probably acts as deterrent to a serious, systematic concern on the part of the teachers with objectives in the affective domain. It is this fast obstacle that this study is focused, with specific attention given to the measurement of attitudes.

During the past four decades a number of instruments have been reported for measuring attitudes relevant in science education, some of which have been used in research studies. Perhaps the best known and most extensively used in the *Test on Understanding Science* (TOUS) developed by Klopfer and Cooley[4]. This test is, in fact, a composite designed to measure understandings about the nature of the scientific enterprise, scientists and the methods and aims of science. A test of measure scientific attitude was developed by Noll[5,6] with focus on the following characteristics: accuracy in operations, intellectual honesty, open-mindedness, suspended judgement, looking for true cause-and-effect relationship, and criticalness. Allen's scale, *Attitudes Toward Science and Scientific Careers*[7], pertains to characteristics of scientists, the nature of scientific work and the contributions of science to mankind. Lowery's *Projective Test of Attitudes*[8] makes use of indirect techniques to measure student attitude toward science, scientists and science processes. A *Science Support Scale* developed by Schwirian[9] is based on Barber's[10] summary of five cultural values which he considers to be conducive to the development of positive scientific attitudes. These five values are rationality (acting on the basis of available evidence), utilitarianism, (interest in natural phenomena), universalism (judging scientists only on the basis of their qualifications), individualism (commitment to individual conscience), and meliorism (acceptance of the benefits of science). *An Inventory of Scientific Attitude* by Moore and Sutman[11] is designed to measure three intellectual attitudes based on some knowledge of the attitude object and three emotional attitudes based on emotional reactions toward science.

Most of the above mentioned instruments are based on a fairly explicit rationale. However, each one suffers from one or more of the following shortcoming: the definitions of attitudes are too general; there is a tendency to lump together several dimensions of science under the caption of attitudes (e.g., interests, attitudes and values are grouped with processes involved in scientific inquiry); the scale do not discriminate between the affective and cognitive components involved in attitude measurement; and the content of the scales often does not adequately represent classroom situations and experiences.

The purpose of the present study was to develop a rationale for evaluating the achievement of attitude objective in science education, in which an attempt is made to eliminate the above mentioned shortcomings. In addition, the practicality of this rational was assessed through application in the preparation and field-testing of an attitude test.

RATIONALE FOR THE STUDY

The main purpose of evaluation in the effective domain should be to guide the development and improvement of teaching methods and rather than to assign pupil marks [1]. Student scores on valid evaluation instruments can be used as an indication of the extent to which the applied methods and materials encourage the desired behaviors. The preparation of these instruments is facilitated if the objectives are defined in behavioral terms.

The objectives which are generally included in the affective domain involve the development of interests, attitudes, values, appreciations, and adjustments[12]. In developing their taxonomy of affective objective, Krathwohl *et al.*[13] did not use the above categories because they felt that the variety of meanings associated with them rendered them inadequate to serve as a basis for the construction of a continuum. Instead, they developed a general classification scheme which could apply to all subjects. There taxonomy defines the affective domain in terms of a valuing system, and involves levels of internalization of values proceeding from simple awareness of stimuli to adherence to an overall philosophy of life. It appears easier to apply this taxonomy in the social science and fine arts that in the natural science, and the problems involved in the one major attempt to apply it to science education[14, 15] underline the difficulty in implementing its hierarchical aspect.

Nay and Crocker[2] developed a rational for defining objectives in the affective domain in which the categories of interests, attitudes, values, appreciations and adjustments are retained. The central thesis of their rationale is that behavioral objectives for science teaching in the affective domain can be derived from a determination of the affective behavior of scientists at work. First, it is necessary to identify affective attributes which are exhibited by scientists in their professional activity. Following this, each of these affective attributes can be characterized or defined in terms of specific behaviors manifested by scientists in their professional work and in relationships with peers. These behavioral definitions of the affective attributes of scientists can be adopted as desirable behaviors for students to attain in the science classroom. That is, the behavioral definitions of the affective attributes of scientists can serve as behavioral objectives for science teaching.

Nay and Crocker[2] compiled an inventory of "affective attributes of scientists"; that is, of a list of interests, attitudes, adjustments, appreciations, and values which scientists are generally expected to demonstrate in their work. It was left to the present study to deal with the matter of behavioral definition of affective attributes and to use these definitions in the development of an evaluation instrument. In view of the complexity of this undertaking and the uncertainty of the outcome, it was decided not to take into account all of the sixty-five attributes in the Nay-Crocker inventory but to consider only attitudes. Nay and Crocker did not make a complete distinction between attitudes and adjustments in their inventory. They identified a set of *operational adjustments*, "behaviors which underlie competence and success in science," and a set of *intellectual adjustments*, "behaviors which are foundational to the scientist's contribution to or acceptance of new scientific knowledge." The term attitude is equated to intellectual adjustments. The eight attitudes which are examined in the present study are critical-mindedness, suspended judgement (restraint), respect for evidence (reliance on fact), honesty, objectivity, willingness to change opinions, open-mindedness, and questioning attitude.

One further element in the rational for this study relates to the nature and dimensionality of attitudinal behavior. Rokeach[16] defines attitudes as a "relatively enduring organization of beliefs around an object or situations predisposing one to respond in some preferential manner." He defines attitudes further in terms of three components: Cognitive, affective, and behavioral. These represent respectively knowledge about, a tendency to take a positive or negative position toward, and some type of observable action with respect to the attitude object or situation.

Nay and Crocker define two components for each affective

attribute listed in their inventory. These correspond to the cognitive and behavioral components defined by Rokeach. The cognitive component represents the student's understanding of the role of the affective attributes in the activities of scientists, while the behavioral component represents the tendency for the student to demonstrate these attribute in his own science work.

In the present study, a multi-dimensional approach incorporating the ideas of Nay and Crocker and Rokeach is used to define three components of each of the attitudes selected from the Nay-Crocker inventory. These are labeled as the "cognitive," "intent," and "action" components. The cognitive component represents the student's understanding of the manner in which attitudes manifest themselves in the professional behavior of a scientist. The intent component represents the student's tendency to show approval or disapproval of behaviors which define an attitude. This is indicated by his endorsement of specific sources of action in certain situations relevant to the attitudes. The actions component represents the extent to which the student actually demonstrates in the science classroom in the behaviors which define an attitude.

The final element in the rational underlying this study relates to the role and specifics of evaluation of affective behavior. The modification of Engman's mode[17] for education planning presented in Figure 1 incorporates evaluation as a check on the effectiveness of the methods and materials used at Phase II in achieving the

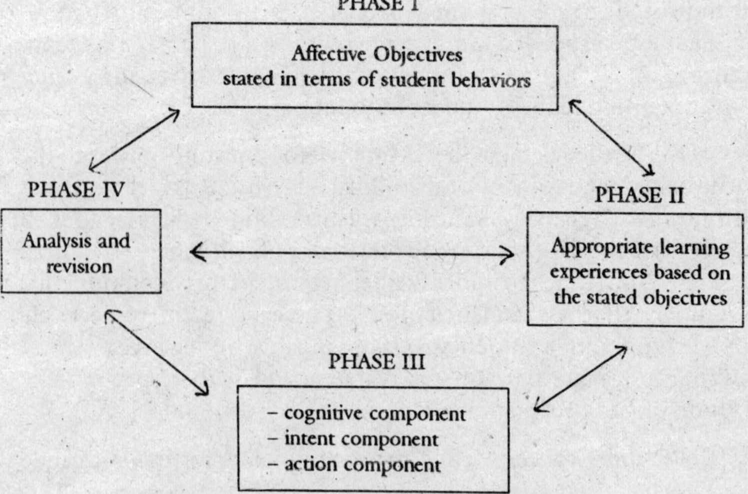

Figure 1 : A model for educational planning.

objectives defined at Phase I. The evaluations provides information to guide the analysis and revision represented at Phase IV. One approach to guiding the instructional and evaluative phases by means of objectives is to state them in terms of observable student behaviors. Consequently, the first step in the study was to determine the behavioral specification of each of eight attitudes listed above, on the basis of which test items were constructed to measure cognitive and intent behaviors. Actually, in this study objectives are stated at two levels, one more general than the other. For example, a general objective is to have students demonstrate "suspended judgement" in science work. However, this type of attitude is demonstrated by more specific behaviors such as the student "generalizes only to the degree justified by the available evidence" or "indicates conclusions as being tentative" (see Table I).

In the present study, Phases II and IV of the model are not investigated. Instead, the assumption is made that regardless of what objectives a teacher has in mind and the activities used in the science classroom to achieve them, students will develop certain attitudes. Therefore, it is possible to identify those attitudes which a given group of students possess without investigating the dynamics of attitude development and change.

THE TEST FORMAT

Science attitude are mediating variables, they cannot be measured directly and must be inferred from some overt response. The most common approach to attitude measurement is to obtain a measure of the respondent's agreement or disagreement with a set of opinion statements about the attitude object or situation. The position taken in the present study is that, since attitudes are defined as predispositions to some preferred response, a reasonable approach to attitude measurement would be to make inferences about an individual's attitudes from his endorsement, or lack of it, of various courses of action in certain situations relevant to the attitude object or situation.

Starting with the work of Thurstone, several modes or strategies have been developed for measuring attitudes. The major ones are the Thurstone scale[18–20], Likert scale[21], and semantic differential[22]. Other methods used in attitude measurement are questionnaires, interview schedules, sentence completion, picture interpretation, word association and error-choice[23–24]. Each of the above methods has drawbacks and limitations. The Thurstone scale are very appealing to use in measuring attitudes, since Thurstone developed a sound theoretical and mathematical foundation to support the analytical procedures which are used in the calculation of the scale values. However, in the construction of his scale, Thurstone assumed a unidimensional attitude object. Therefore, a large number of scales would be required to identify all of the dimensions of the affective domain in science education. Since a considerable amount of work is required on the part of the respondents who provide the data from which the scale is to be determined, the construction of large number of scales may not be practical undertaking. A major weakness with both the Likert sales and semantic a differential is the possibility of response bias on the part of the respondents, such as tendency to choose extremes[25-26]. In addition, the semantic differential technique was deemed inappropriate for the present study because the information obtained is not directly related to the situation in a science classroom.

As a result of a critical survey of the reported techniques for measuring attitude. It was decided to use the multiple-choice item for testing the cognition and intent components of attitudinal behavior. The multiple-choice item includes a stem describing a situation relevant to a given attitude and distractors describing different courses of action. This is constant with the position taken in this study that an individual's attitude can be inferred from his endorsement of certain courses of action relevant to the attitude object or situation.

PROCEDURE

The procedure consisted of the following stages; behavioral definition of each of the eight attitudes mentioned above, construction of a test, administration of the test to students, and analysis of the test results.

An initial list of behavioral objectives was compiled for each of the eight attitudes. Appropriate objectives found in the literature were included in this list[15, 27, 28]. A panel consisting and graduate students in science education rated these objectives on a 0-1-2 scale for relevance to a specific attitude. This initial list was reduced to less than half on the basis of an arbitrary cutoff on total points received by each objective. The final list of behaviors defining each attitude is presented in Table 1. It should be noted that the behaviors defining critical-mindedness and questioning attitude are identical. Consequently, henceforth both of these attitudes will be referred to as critical-mindedness. Similarly, open-mindedness will be sub sumed under objectivity.

Multiple-choice questions were constructed to reflect the behaviors in Table 1. A large amount of science testing material was surveyed in a attempt to find suitable questions. This search was not very productive, although some materials did provide ideas for situation on which question were based[15, 27, 29, 30].

A pilot study was conducted of the first draft of the test which consisted of 28 items. On the basis of the results from this study and student comments, some of the items were revised and twelve more were prepared to make up the forty items in the *Test on Scientific*

Table I
Behavioral Description of Attitudes

A student demonstrates critical-mindedness when he:

- looks for inconsistencies in statements and conclusions.
- consults a number of authorities when seeking information.
- looks for empirical evidence to support or contradict explanations.
- asks many questions starting with what, where, why when and how.
- challenges the validity of unsupported statements.

A student demonstrates suspended judgement (restraint) when he:

- generalizes only to the degree justified by available evidence.
- collects as much data as possible before drawing conclusions.
- recognizes conclusions as being tentative.
- consults several authorities (texts, periodicals, people) before drawing conclusions.

A student demonstrates respect for evidence (reliance on fact) when he:

- looks for empirical evidence to support or contradict explanations.
- collects as much datá as possible before drawing conclusions.
- demands that explanations fit the facts.
- demands supportive evidence for the unsubstantiated statements.

(Contd.)

Table I (contd.)

- supplies empirical evidence to support his statements.

A student demonstrates honesty when he:

- reports observations even when they contradict his hypotheses.
- acknowledges work done by others.
- considers all available information when forming generalization and drawing conclusions.

A student demonstrates objectivity when he:

- considers all available data(not only that portion which supports his prior hypotheses).
- reports observations even when they contradict his hypotheses.
- considers and evaluates ideas presented by others.
- examines many sides of a problem and considers several possible solutions.

A student demonstrates willingness to change opinion when he:

- recognizes conclusions as being tentative.
- recognizes that knowledge is incomplete.
- considers and evaluates ideas presented by others.
- evaluates evidence which contradicts his hypotheses.
- alters his hypotheses when necessary to accommodate empirical data.

A student demonstrates open-minded when he:

- considers and evaluates ideas presented by others.
- evaluates evidence which contradicts his hypotheses.
- considers several possible options when investigating a problem.
- considers both pros and cons when evaluating a situation.

A student demonstrates questioning attitude when he:

- looks for inconsistencies in a statements and conclusions.
- consults a number of authorities when seeking information.
- looks for empirical to support or contradict explanations.
- asks many questions starting with who, where, why, when and how.
- challenges the validity of unsupported statements.

Attitudes (TOSA) shown in the Appendix. A panel of judges reacted to the keyed responses for the test items.

TOSA is divided into two sub-tests of twenty items each, which were randomized in the administered test. The stems of items 1-20, the *Cognitive Component Sub-test* (CCS), describe a situation which a scientist might encounter in his work. The student is asked to select from four courses of action the one which is most appropriate for the scientist. The items 21-40, the *Intent Component Sub-test* (ICS), present a situation which the student may encounter in the science classroom or in everyday activities. The students is asked to select from four alternate course of action the one which best represent his intent reaction to this situation.

The classification of the multiple-choice question given in Table II was made on the basis of the apparent relationships between the behavioral definitions of the attitudes and the alternatives in the questions. When the items were written an attempt was made to distribute them evenly among the six attitude and between the CCS and ICS sub-tests. Since several of the behaviors are listed under more than one attitude. When constructing related to the same attitude. However, this was not possible for most questions. This factor also contributed to the uneven distribution of items among the attitudes.

TOSA was a administered to 307 students in grade eleven physics and chemistry classes. The test was re-administered to 105 of these students three weeks after the first testing to provide test-re-test stability data. Three groups (numbering 38, 51 and 62 students, respectively) were rated by their teacher, on the extent to which each student exhibit the behaviors defining the attitudes involved in this study. This rating was on a four-point scale for which 4 indicated frequent demonstration of these behaviors. Only one rating on the overall set of six attitudes was obtained for each

Table II
Classification of Test Items

ATTITUDE	COGNITION	INTENT
Critical-mindedness (questioning attitude)	9, 19	21,24,25,31,32,36
Suspended judgement (restraint)	1,2,4,6,7,11,18	24,26,27,28,34,35,37
Respect for evidence (reliance on fact)	10,11,12,14,16	26,27,28,31,32,33,38
Honesty	3, 8, 18	22,33,39
Objectivity (open-mindedness)	8,13,15,17	23,29,30,36,40
Willingness to change opinions	1,3,4,5,6,10	29,30,37

student. These were the only data were obtained on the "action" component of attitudinal behavior.

Because the sample was restricted to students taking grade eleven chemistry and physics the present study has limited value with respect to generalizability of the results. However, since this is basically a methodological study, the main question of interest in whether or not the constructed test items adequately identify selected characteristic of the student in the sample.

In the analysis of the data a number of descriptive statistics are derived for the test and sub-tests. Due to a misprint in the test, item 20 was not used in the analysis. Item analysis is used to study the properties of the individual items, and factor analysis is used to examine their underlying structure.

RESULTS

Descriptive Statistics: For the 307 students tested, the means for TOSA, CCS and ICS are 52.4, 52.8 and 52.2 percent, respectively. The standard deviations are 103. 13.9 and 12.5, respectively.

The correlation between CCS and ICS is 0.23. This lends support to the division of the questions into the two subjects since this correlation is considerably lower than the odd-even, split-half correlation for TOSA (0.40).

Since many of the questions in TOSA require a considerable amount of reading, correlations between the test scores and reading ability were calculated. Scores on the *Cooperative Sequential Test of Educational Progress* (STEP) in reading, form 3A or 3B, were obtained for 248 students. The correlation coefficient for TOSA, CCS and ICS with STEP reading are 0.35, 041 and 0.13, respectively. Although these coefficient are significantly different from zero at the 0..01, and 0.01, and 0.05 probability levels, respectively, they do not indicate a strong relationship between reading ability and scores on TOSA, CCS and ICS.

Item Analysis: Students were allowed sufficient time to complete the test and all of the students responded to all of the questions. Most of the alternatives proved to be acceptable distractors. Out of the 156 alternatives analyzed for the thirty-nine questions (items 20 was not included), seven alternative received less than three percent of the responses and none of the distractors was completely ignored. Most of the distractors that received one to two percent of the responses are in questions with difficulty index of 0.80 or higher. It is possible that the nature of the sample which was tested may have physics to raising the difficulty index of some of the easier questions. Chemistry and physics are not compulsory courses; therefore, most of the students registered in these causes will be taking the course because of their interest in the subject. Also, a certain amount of the screening done before students are allowed to take these courses.

The item difficulties and biserial correlation coefficients are given in Table III. Biserial correlations for each test item with the total test score (TOSA) with the two sub test scores (CCS and ICS) are included.

The item difficulties range from 0.11 to 0.87. Seven questions have difficulty levels higher then 0.75 and five have difficulty levels lower than 0.25.

Some bi-serial correlations are quite items with the sub-test scores will be spuriously high because the sub-tests are quire short, hence unreliable. However, most of these coefficients are near 0.30 and should be satisfactory. Question 19 and 22 have coefficients not significantly different from zero, hence should probably be revised or remitted from the test.

Reliability. The KR-20 coefficient is used as a measure of homogeneity, and the test-retest correlation coefficient is used as a measure of test stability. For the sample of 207 students, the KR-20 coefficient for TOSA, CCS and ICs, and 0.55 0.45, and 0.39, respectively. A number of factors may have contributed to these low coefficeints. There is some evidence that some of the questions are not strongly related to the rest of the test that the results of the item and factor analyses. In the construction of the test, no attempt was made to ensure that the item difficulties would be near 0.5. The fact that twelve of the questions have difficulty indices above 0.75 or below 0.25 will tend to lower the KR-20 coefficients. The coefficients will also be affected by the test length and studied because the sample employed was homogeneous[31].

The test retest correlation coefficients for TOSA, CCS, and ICS are 0.71, 0.68, and 0.64 respectively. These correlations were calculated from scores for 105 students for a test-retest period of three weeks.

Table III
Difficulty Levels and Biserial Correlations for Test Items in TOSA

ITEM	DIFFICULTY*	BISERIAL CORELATIONS		
		TOSA	CCS	ICS
1	0.85	0.22	0.29	
2	0.61	0.37	0.36	
3	0.59	0.34	0.49	
4	0.52	0.27	0.27	
5	0.66	0.40	0.46	
6	0.55	0.44	0.46	
7	0.52	0.47	0.54	
8	0.32	0.23	0.37	
9	0.82	0.41	0.43	
10	0.80	0.60	0.60	
11	0.54	0.22	0.25	
12	0.55	0.33	0.42	
13	0.22	0.29	0.41	
14	0.40	0.21	0.35	
15	0.26	0.25	0.38	
16	0.50	0.30	0.37	
17	0.58	0.42	0.57	
18	0.42	0.41	0.52	
19	0.18	0.01	0.08	
21	0.20	0.24		0.32
22	0.11	0.07		0.24
23	0.37	0.27		0.35
24	0.64	0.27		0.30
25	0.35	0.21		0.30
26	0.87	0.65		0.63
27	0.58	0.32		0.38
28	0.68	0.36		0.47
29	0.60	0.38		0.46
30	0.49	0.35		0.38
31	0.81	0.34		0.51
32	0.86	0.48		0.62
33	0.85	0.51		0.44
34	0.39	0.37		0.41
35	0.50	0.34		0.38
36	0.18	0.25		0.29
37	0.39	0.27		0.41
38	0.56	0.19		0.33
39	0.56	0.20		0.27
40	0.34	0.15		0.34

* Item difficulty is given as the percentage of students selecting the keyed response.

Validity: The validity of TOSA, CCS and ICS is examined under the three components of validity (substantive, structural, and external) defined by Loevinger[32]. These three components incorporate the concepts of content, construct, predictive, and concurrent validity discussed by other writers[33,34].

1. The *content validity* (substantive component) can be argued on the basis that the attitudes which the test is designed to measure were selected from a list of affective attributes of scientists; the behavioral specifications of these attitudes were selected on the basis of responses of a panel of judge; the content of the items describe science-related situations; and the content of the items is comparable with the idea expressed in a wide variety of science reading material. The validity of the keyed responses has been demonstrated by a panel of judges. These points have been discussed in more details earlier in this paper.

2. Factor analysis was used to examine the *structural validity* of TOSA. The empirical structure underlying the test items, as identified by a factor analysis solution, is compared with the structure predicted (see Table II) from the behavioral definitions of the attitudes in Table I.

The inter-item correlations for TOSA (excluding item 20) were examined by common factor analysis, and the nine factors identified in Table IV were retained. The factor analysis techniques used in this study made use of an oblique rotational procedure developed by Harris and Kaiser[35]. The rotation was performed on a loading matrix obtained by the unweighted least squares factoring of the inter-item tetrachoric correlation matrix. Items 14,19,30 and 40 are not included in any of the factors. All of the these items except 19 have very low communalities. This lack of relationship with the rest of the test is also indicated by low bi-serial correlations for items 14,19, and 40 (see Table III). Therefore, it is not likely that these items would show high loading if additional factors were obtained.

The reported factor solution lends some support to the division of the items into the Cognitive Component Subtest and Intent Component Subtest. As can be seen from Table IV, four of the factors consist mainly of items from ICS and two factors consist mainly of items from CCS.

In the following discussion an attempt is made to relate the

Table IV
List of Test Items in Each Factor

Factor	Cognition	Intent	Attitude
I	6	22, 24, 26, 32	suspended judgement
II		21, 22, 28, 29, 31, 32, 33, 34	respect for evidence
III		23, 29, 36	objectivity
IV	8, 15, 17, 18	33, 35, 37	objectivity
V	2,3,5,6,7,10,12,17		willingness to change opinions
VI	5	25,32,39	critical-mindedness
VII	4,13,18	26,27	suspended judgement
VIII	1,3	32	willingness to change opinion
IX	9,10,11,16	26,33,38	respect for evidence

grouping of the test questions by the factors analysis (Table IV) to the initial classification (Table II) based on the behavioral definition of the attitudes. The factor names given in Table IV were selected by examining the questions involved in each given factor to determine the attitude to which the majority of them are related.

Questions 6, 24 in *factor I* are listed under *Suspended judgement* in Table II. The keyed response for question 32 (see Appendix) is related to the critical-mindedness and respect for evidence. However, the decision not to select the other distractors demonstrates suspended judgement. Question 22 which deals with interpretation of data not appear to have a great deal in common with the other questions in this factor. It is somewhat process oriented.

Question 28, 31, 32 and 33 in *factor II* are listed under *Respect for evidence* in Table II. Question 21 (questioning beliefs because scientists have cast doubt on some of them) appears to be more related to critical-mindedness, but a respect for evidence is also implied in this response. Question 22,29 and 34 do not appear to be directly related to respect for evidence.

All of the question in *factor III* (23, 29, and 36) are listed under *objectivity* in Table II.

Question 8, 15 and 17 in *factor IV* are listed under *objectivity* in Table II. The keyed response to question 33 (reporting results which do not support expectation) relates to objectivity as well as honesty. The keyed response to question 35 reflects willingness to consider contradictory hypotheses, which is also characteristics objectivity. Question 18, 35 and 37 in this factor have been classified in Table II and under-*Suspended judgment*. Factor IV was labeled *Objectivity* because the factor loadings for questions 8, 15 and 17 are higher than the factor loadings for questions 18, 35 and 37.

Question 3, 5, 6, and 10 in *factor V* are listed in Table 2 under *Willingness to change opinions*. Question 2 and 7 are more related to *Suspended judgment*, while question 12 and 17 are more related to *Respect for evidence* and Objectivity, respectively.

Question 25 and 32 in *factor VI* are listed under *Critical-mindedness* in Table II Priestly's behavior as described in question 5 demonstrates unwillingness to change opinions and a lack of respect for evidence. However, he also failed to demonstrate critical-mindedness in his evaluation of the phlogiston theory. Question 39 does not appear to be related to critical- mindedness.

Question 4, 18, 26 and 27 in *factor VII* are listed under *Suspended judgment* in Table II. The situation described in question 13 (reasons why Arrhenius theory of ionization was not widely accepted) is more relevant to *Objectivity* than to *Suspended judgement*. However, Arrhenius demonstrated suspended judgement in his search for a new theory.

Questions 1 and 3 in *factor VIII* are listed under *Willingness to change opinions* Table II. Questions 32 appears to be more related to *Critical-mindedness* and *Respect for evidence* than it is to *Willingness to change opinions*.

Questions 10, 11, 16, 26, 33 and 38 in *factor IX* are listed under *Respect for evidence* in Table II. Question 9 is classified under *Critical-mindedness*. However, the philosophers' actions appear to reflect a lack of respect for evidence.

All of the attitudes appear in Table IV expect the attitude of honesty. Honesty has not been associated with any of the factors, possibly because it is not defined very distinctively. Two of the behavioral objectives defining honesty are very similar to behaviors related to the evaluation and reporting of data which is contradictory to predicted hypotheses. The other behavioral objective defining honesty (acknowledges work done by others) is not easily translated into a suitable test question.

Some of the attitudes are associated with two factors. In these cases, one factor is generally a mixed factor containing questions from both the CCS and ICS sub-tests, while the other factor is composed mainly of questions from only one of the subtests. In the light of the above evidence from factor analysis it seems inappropriate to divide TOSA into subsets on the *basis of the attitudes*. The main reason for this is that the overlap of behavioral objectives used to define the attitudes militates against a distinct division of the test items. In the factor solution presented above, ten of the questions are included in more than one factor.

The purpose of the factor analysis is not to provide sub-tests but to substantiate the classification of the test items on the basis of the rationale used for this study. The agreement between the classification presented in Table II and IV is reasonably good, in that thirty-one test items relate to the same attitude in both tables. In classifying the items on the basis of the rational, it is possible to anticipate all of the factors which might influence a respondent's decision to select a particular response. The items were classified on the basis of the more obvious relationships with the behavioral objective use to defined the attitudes. In an attempt to interpret the factor solution, less obvious relationships were revealed for seven of the test items. Since there are fifty entries in Table IV, it is possible to relate approximately 75% of the salient factor loadings to the item classification based on the rationale used for this study.

3. It was hoped to examine the *external (concurrent) validity* of TOSA on the basis of teacher ratings of the which each student exhibited the behaviors which were used to define the attitudes. The method of obtaining teacher rating is discussed under the procedure. From these ratings a high student group and a low student group could be identified the performance of these two groups on TOSA and its two subsets could be compared. However, this was not done as same doubt arose regarding the validity of the teacher ratings.

Correlations are given in Table V for the teacher ratings of students with their scores on TOSA, CCS and ICS and the final

Table V
Correlations of Test Scores with Teacher Ratings

Test	Group Size 62	51	38
TOSA	0.43**	0.20	0.08
CCS	0.28*	0.10	0.04
ICS	0.39**	0.23	0.09
FSM	0.41**	0.67**	0.33*

* Significantly different from 0.0 at the 0.05 probability level.
**Significantly different from 0.0 at the 0.01 probability level.

science marks (FSM) which were assigned by the teachers at the end of the school term. The correlations were calculated for each group separately because it felt that each of the three teachers may have understood the rating operation differently. The most consistent correlation across the three groups is the correlation of teacher rating with the final science mark. It appears that student achievement in science may have had a major influence on the teacher ratings. Therefore, the teacher ratings of student attitudinal behavior may be of somewhat questionable validity and are probably inappropriate to use a criterion to examine the external validity of TOSA. It is possible, however, that the cognitive and intent components do not correlate highly with the action component. That is, a student's behavior in a given situations may vary considerably from his express intentions with respect to that or some similar situation.

The information obtained in the present study does not account for the wide variation in correlations for the three teachers. It is possible that the teachers may have interpreted the instructions differently. It is also possible that the three teacher may not have been equally aware of the affective development of their students. Since the group sizes are quite small, some of this variation will be due to sampling error.

CONCLUSION

The rationale outlined above provided valuable guidance for the construction of the test questions in the *Test on Scientific Attitudes.* In addition, on the basis of the support provided by the factor analysis of the test results, it can be concluded that this rationale is a useful guide for the division of the test items into the cognitive and intent sub-tests and their classifications on the basis of the behavioral definitions of the attitudes.

The present study indicate that the cognitive and intent sub-tests are not measuring the same characteristics. Student understanding of how scientists demonstrate attitudes in their work probably is not sufficient to ensure that students will demonstrate these characteristics in their own science work or in everyday activity.

Obviously, further research is called for on every aspect of this approach for evaluation of attitudinal behavior of students. There is a need for a better congruence of the application of the rationale with results from test analysis. This will require strengthening of the rationale as well as improvement of test items and lengthening of the test. More attention must be given to evaluation of the action component. Finally there is a need to increase the generalizability of the rationale by application to a broader range of grades, subjects, student abilities, and other areas of affective behavior.

APPENDIX

*Test on Scientific Attitude**

Cognitive Component Subtest

1. Scientists recognize that a scientific theory
   - A. should not be changed when it is based on a large amount of data.
   - B. may have to be changed to keep us with a rapidly changing world.
   - C. may have to be changed when new observations are made.
   - D. should not be changed when it explains what happen to nature.
2. A science magazine reports that a scientist produced type of water that boils at 450° F under one atmosphere of pressure.

* Students were to indicate the one BEST answer for each question

Another scientist reading this report would probably

A. believe the report if it was written by a highly respected scientist.

B. disbelieve the report because he would know that water boils at 212°F under one atmosphere of pressure.

C. do experiments to try to prove that it is wrong.

D. neither believe nor disbelieve the report until other scientists study this problem.

3. When observations are made that do not fit an accepted scientific theory, scientific usually.

A. try to adjust the observations so that they fit into the theory.

B. keep the theory as it is since the new observations cannot be use to improve it.

C. try to change the theory so that these observations can be explained.

D. discard this theory and develop a new one to explain these observations.

4. When Einstein published his theory of relativity another famous scientist was reported to have said, "Dr. Einstein's new theory has shattered many of my scientific beliefs to smithereens." This statement indicates that the scientist

A. recognized that scientific knowledge is subject to change.

B. held some wrong scientific beliefs without knowing it.

C. did not believe in the old theory very strongly.

D. did not have sufficient evidence to support his original beliefs.

Questions 5 and 6 refer to the following paragraph :

Priestly and Lavoisier are often referred to as the "fathers of modern chemistry" Both of them accepted the Phiogiston theory of combustion (all materials give off a substance called phlogiston when they burn). However, Lavosier did many experiments as burning and developed our modern theory of combustion in which he said that oxygen is always involved. Preistley never selected this theory.

5. Which one of the following is generally true about scientists, but was <u>NOT</u> demonstrated by Prisetley in the above situation?
   A. Some scientist believe more strongly in their theories.
   B. Some scientists go overboard in demanding experimental evidence before changing their ideas.
   C. Scientists do not have to belief in new theories.
   D. Scientists accept new theories when they are consistent with experimental data.
6. Which one of the following is <u>NOT</u> true about Lavoister in the above situation?
   A. He believed that his theory of combustion would not be changed.
   B. He recognized that theories are likely to change.
   C. He was prepared to consider ideas presented by others.
   D. He developed a new theory to explain new evidence.

Question 7 and 8 refer to the following paragraph :

The German scientific, Schleiden, published a report on the origin of plant cells (1838). He made several observations on the reproductive cells of some plants and made the following statements:

> It is an absolute law that every cell takes its origin as very small vesicle [small bladder] and grows slowly to its defined size. The process of cell formation which I have just described . . . is that process which I was able to follow in most of the plant which I have studied. Yet many modifications remains incontestable [cannot be questioned] . . . .(30).

7. Which one of the following is generally true about scientific but was NOT demonstrated by Schleiden in the above situation?
   A. Scientists try to avoid making general statements based on limited data.
   B. Scientists are usually careful to report exactly what they observe.
   C. Scientists collect large amounts of data in order to develop laws of nature.

D. Scientists often ignore observation if they do not quite fit into their theories.

8. Some aspects of Schleiden's theory were latter shown to inaccurate. The most probable reason why his theory was NOT completely accurate is that he

   A. was not able to obtain modern instruments to use in his investigation.

   B. did not make his theory explain all of his observation.

   C. tried to develop a theory to explain the origin of all cells.

   D. felts that his theory could not be questioned.

Question 9 and 10 refer to the following paragraph:

> Galileo gathered much evidence on stars, motion of objects, etc. which gave rise to ideas contrary to those held by the philosophers of his time. The philosophers forced Galileo to recant some of these ideas (say he was wrong) and stopped hismfrom practicing science.

9. Which one of the following best applies to this situation:

   A. Galileo should have collected more evidence before disagreeing with the philosophers.

   B. Galileo's ideas became wrong when he recanted.

   C. Galileo's should have avoided those investigation which led to disagreement with the philosophers.

   D. Galileo was justified in questioning the beliefs of the philosophers.

10. In their treatment of Galileo, the philosophers.

   A. showed that they did not have a proper respect for evidence.

   B. seemed to think that they knew all that there was to know.

   C. were not willing to change their ideas in the facts of new evidence.

   D. showed all of the above characteristics.

11. Drs. Brown, Jones, and Smith are medical researchers. Each one independently investigated the cancer-producing effect of compounds in tar on rats. Dr. Brown reported that there was

no effect. Some time later, both Drs. Jones and Smith reported that these compounds were highly cancer-producing. Which one of the following was probably the MOST important reason for Dr. Brown's results?

A. He did not consider all the evidence.

B. He did not do a sufficient number of controlled experiments.

C. He was in a hurry to report his results first.

D. He did not analyze his data properly.

12. If a scientist had no choose between two theories, he would probably support the theory which;

A. most other modern scientists feel is more likely to be correct.

B. has more practical value.

C. is based on a larger number of observations.

D. explains the available observations more satisfactorily.

13. When Arrhenius first proposed his theory of ionization (salts break up into ions when they dissolve in water), very few scientists were willing to support it. Which one of the following is the MOST probable reason for this disagreement?

A. Arrhenius gave a different interpretation to the observations related to this problem.

B. The scientists who would not support this theory were not as imaginative as Arrhenius.

C. Arrhenius did not have enough evidence to support his theory.

D. The scientists who would not support this theory were less willing to risk criticism.

14. A scientist was studying an ore from the moon in an attempt to obtain a new metal from it. He made several tests but did not find evidence of a new metal. However, he did identify a peculiar gas which he obtained during one of the tests. He probably would have

A. reported that the ore did not contain a new metal.

B. reported that portion of his investigation related to the gas.

C. not made any report because he did not solve his problem.

D. not made any report until he was able to get another scientist to confirm his identification of the gas.

15. Quite often it is possible to give several different explanations for a particular set of observations. Which one of the following would <u>NOT</u> be generally true about such explanations?

<u>A</u>. Only one of these explanations could be the true scientific explanation.

B. All other things being equal, the explanation which is the most widely known is likely to be accepted one.

C. The explanation which suggests the greatest possibility for further study is likely to be the one which most scientists use.

D. All these explanations would be acceptable if they explain the observations.

16. Quite often two groups of scientists will support opposing theories about some aspect of science. Which one of the following would be the <u>MOST</u> important point to consider in settling such a controversy?

A. Both theories give satisfactory explanations for the observations related to the problem, but one theory has more practical applications.

B. One group of scientists believe more strongly in their theory.

C. One group contains several scientists who have won the Nobel Prize for science.

<u>D</u>. Different conclusions are reached when the two theories are applied to certain problem.

17. A scientist shows that he is open-minded when he

A. discusses his ideas with other scientists.

<u>B</u>. evaluates ideas which do not agree with his theories.

C. agrees with the ideas presented by other scientists.

D. asks other scientists to provide experimental evidence to support their arguments.

18. Theories in science are generally accepted when it can be shown that they explain all of the observations. However, it is possible that exceptions to the theory may exist but are still undiscovered. Which one of the following is the BEST approach to this problem ?

    A. The limits under which the theory has been shown to apply should be carefully stated and the theory should be used within these limits.

    B. Scientists should provide several theories to explain a given set of observations so that if exceptions to one theory are found, they will have others to rely on.

    C. Scientist should not accept a theory until they are certain that exceptions to it do not exist.

    D. When exceptions are discovered, scientists should abandon the theory and look for a new one.

19. Which one of the following is NOT an important reason why scientists often repeat the experiments reported by other scientists?

    A. A scientists could be so intent on finding a specific answer that he might subconsciously observe only what he wants to see in his experiments.

    B. This helps to keep scientists careful and honest when making observations and reporting results.

    C. Other scientists might give a different interpretation to the same observations.

    D. The first scientist might overlook a significant variable in this experiment.

20. A scientist has a theory for which he needs some evidence. He does experiments and finds that some of the results do not support his theory. When he reports his theory he omits those results which do NOT fit. In this case, the scientist

    A. had a theory which did not have any practical value.

    B. considered several possible explanations.

    C. made his theory explain parts of the experimental results.

    D. made the experimental results agree with his theory.

Intent Component Subtest

21. Scientists have questioned many religious beliefs. Which one of the following best expresses the way you feel concerning this matter?

    A. When scientific theories question religious beliefs, it is better to keep the religious beliefs.

    B. I now question all of my religious beliefs since science has cast doubt on some of them.

    C. I have two separate thought compartments (one for my religious beliefs and one for scientific knowledge).

    D. I will keep my religious beliefs until scientists prove them to be wrong.

22. Imagine that you have just finished a laboratory investigation. Your measurements all agree except two. Which of the following would you do?

    A. Include the two add measurements in your reports but omit them from calculations.

    B. Adjust the two odd measurement to make them agree better with the others.

    C. Take more measurements.

    D. Use all the measurements as they are when doing calculations.

23. Consider the following data concerning fluoridation of the public water supply:

    Fluorides help prevent cavities in children's teeth but do not help adult teeth.

    Small amounts of fluorides appear to have no long-term harmful effects.

    The earliest and cheapest way to administer fluorides is through the public water supply.

    The fluoride content of lakes and oceans is increasing as a result of fluorides in the public water supply.

    Fluorides can be put in milk for children.

    Which one of the following best describes your point of view after considering the above information?

    A. You would be against fluoridation.

B. You would be uncertain as to which side to support.

C. You would be in favor of fluoridation.

D. You would loss interest in the problem because the evidence is too indefinite.

24. "Light travels as a stream of particles."

"Light travels as a wave."

If you come across these two statements in two different science books, which of the following would you do?

A. Ask your teacher to tell you which statement to accept.

B. Check other science books for statements on this topic.

C. Assume that scientists are not certain as to how light travels.

D. Accept the statement in the newer book.

25. Imagine you are living in a small town on the banks of a river not far from a large industrial city. Your town has just experienced a sever flood for the first time in its history. Some people are saying that it was caused by increased rainfall due to the smog from the nearby industry. Which one of the following best expresses your evaluation of this claim?

A. This is a popular opinion for which there is no evidence.

B. People are making this claim because of their prejudice against smog.

C. This is a valid conclusion based on sufficient evidence.

D. This is a popular opinion backed by some evidence.

26. Suppose that you and a friend both did the same experiment to determine whether or not sunlight is required for plants to produce starch. Both of you tested a leaf from a plant that had been left in the dark for two days. Your friend found starch in both leaves. You found starch only in the leaf from the plant that had been left in the sunlight. Which one of the following would be the most reasonable thing for you to do?

A. Accept your own result because text books say that plants in the dark should not produce starch.

B. Have both of you repeat the experiment.

C. Accept the result obtained by the one of you who knows more about science.

D. Ask your teacher to decide which result should be accepted.

27. Suppose you wanted to determine which type of mosquitoes cause malaria. You obtained there kinds (Types A, B, and C) and examined the digestive tracts of each for malaria parasites. You found some only in Type B mosquitoes. You concluded from this that malaria is spread by type B but not by Type A and c mosquitoes. Which of the following describes your conclusion?

A. Your conclusion does not agree with the evidence.

B. Your conclusion is valid in light of the evidence.

C. Your conclusion is justified, but more evidence should be obtained.

D. You did not obtain enough evidence to make a conclusion.

28. Some medical researchers say that marijuana does permanent damage to the brain, while others say that it is more harmful alcohol. In the light of this information, which of the following would you be inclined to do?

A. Not smoke it because it is probably harmful.

B. Ignore the evidence that it might be harmful and smoke it if you wanted to.

C. Smoke it because it is probably no more harmful than alcohol.

D. Put off any decision about smoking it until more definite knowledge is obtained about its effects.

29. "Many people have cycles of mental depression which correspond to the phases of the moon." Which one of the best represents your reaction to this statement?

A. One should be willing to consider the possibility that there may be some truth to superstitious of this nature.

B. Scientists could never prove or disprove this idea.

C. It is an incorrect data, but is useful to many people.

D. There seems to be some truth in this statement.

30. Below are a number of points of view regarding the teaching of the theory of evolution in biology. In your opinion, this theory should be

    A. omitted from the biology course.

    B. presented to the class, but its controversial aspects should not be discussed.

    C. discussed thoroughly in class with all students present.

    D. discussed openly in class, but those students who do not want to listen should be permitted to leave.

31. Suppose you live near a large industrial plant. You find that the rose bushes in your yard die in a short while, but lawn remains in perfect condition. You suspect that the fumes from the industrial plant are the cause. Which one of the following would be the most reasonable course of actions for you to take?

    A. Study the effect of the fumes on healthy rose bushes.

    B. Stop growing rose bushes.

    C. Start legal action against the plant for pollution control.

    D. Move away from the plant.

32. During a class discussion, a friend of yours said, "The questions which the really important to man can never be solved by science." Which one of the following would probably be your reactions to this statements?

    A. Support him because friends should stick together.

    B. Not pay any attention to this statement because it is not worth thinking about.

    C. Ask him to present facts and argument to support his statement.

    D. Support his because you believe the statement is true.

33. Suppose you did a chemistry experiment, but the results were not what you expected. Which one of the following would you do ?

    A. Report the results which were predicted in the chemistry text.

B. Copy the results from a friend.

C. Report the results that you obtained.

D. Report no results and tell the teacher that the experiment failed.

34. A boy goes skating on a pond and breaks through the ice. He is rescued and given a drink of hot chocolate by someone who is sneezing and caughing. A few days later the boy also has a cold. Which one of the following best describes the reason for the boy's cold?

A. His cold is due to falling in the water and getting wet.

B. He got the cold from the person who rescued him.

C. He probably had a cold coming before he went skating.

D. The reason why people get colds is not yet known for certain.

35. In an experiment students blew through limewater and noted that it turned milky. From this result most of them concluded that their bodies give off carbon dioxide. However, one girl wrote in her notebook that since there is carbon dioxide in the air we breathe, the experiment proved nothing. Which one of the following best describes your evaluation of this situation?

A. The students were justified in making their conclusions.

B. The girl was justified in doubting the proof.

C. Neither side had sufficient grounds for their statements.

D. Both sides were partly justified in their statements.

36. "People born when certain stars are becoming more prominent show the influence of these stars in their personalities." People who believe this statement

A. probably have a special ability to understand such influences.

B. are not critical enough.

C. are more open-minded than most people.

D. have disregard for scientific evidence.

37. When evaluating the accuracy of ideas in science texts, which one of the following is the most important?

A. How recently the book was published.

B. Whether or not the author is a scientist.

C. The extent to which the ideas have been simplified.

D. How recently the ideas were first presented.

38. If you came across a scientific idea which goes against your common sense, which one of the following would you be inclined to do?

A. Disregard the scientific ideas because it is better to rely on common sense.

B. Disregard common sense because it is not as reliable as scientific study.

C. Do an experiment to see whether or not the common sense is superior to the scientific idea.

D. Try to produce compromise between the scientific idea and common sense.

39. Suppose you had worked several day on a chemistry experiment. You then accidently added some sodium nitrate solution when you should have added silver nitrate. Which one of the following courses of action would you take?

A. Start over again as soon as you realize your mistake.

B. Continue with the experiment but if it doesn't turn out the way it should, start over.

C. Continue the experiment to see if the mistake makes any difference.

D. As soon as you realize your mistake, add some silver nitrate solution and continue with the experiment.

40. A missionary reported that the root of a plant much like the Rauwolfia plant had been used by African witch doctor to cure him of a serious illness. Recent medical reports show that reserpine, a drug effective in lowering blood pressure, is extracted from Rauwolfia. Which one of the following is the most reasonable conclusion that can be drawn from the above discussion?

A. Since the witch doctor probably did not know anything about modern drugs, he did not have scientific reason for using the roots.

B. The plant was probably not helpful because the missionary had no way of knowing what caused him to get better.

C. The plant may have been helpful since the missionary recovered after the witch doctor's treatment.

D. The plant probably was helpful because the Rauwolfia plant contains reserpine.

•

References

1. Bloom, B.S., J.T. Hastings, and F.G. Madaus, *Handbook on Formative and Sammative Evaluation of Student Learning.* New York, McGraw-Hill Book Co., 1971.

2. Nay, M.A. and R.K. Crocker, "Science Teaching and Affective Attitude of Scientist," *Science Education*, 54; 59-67, (1970).

3. Scriven, M., "Student Values as Educational Objectives." In *Proceedings of the 1965 Invitational Conference On Testing Problems.* Princeton, Educational Testing Service, 1966.

4. Klopfer, L.E., and W.W. Cooley, *TOUS: Test on Understanding Science*, Form W, Princeton, Educational Testing Service, 1961.

5. Noll, V. H., "The Habit of Scientific Thinking," *Teachers College Record*, 35: 1-9, (1933).

6. Noll, V.H., "Measuring the Scientific Attitude, " the *Journal of Abnormal and social psychology*, 30: 145-154, (1935).

7. Allen, H., Jr., *Attitudes of Certain High School Seniors Towards Science and Scientific Careers*, New York, Marstin Press, Inc., 1959.

8. Lowery, L.F., "Development of An Attitude Measuring Instrument for Science Education," *School Science and Mathematics*, 66: 494-502, (1966).

9. Schwirian, P.M., "On Measuring Attitudes Toward Science," *Science Education.* 57: 172–179, March 1968.

10. Barber, B., *Science and the Social Order*, New York, Collier Books, 1962.

11. Moore, R.W., and F.X. Sutman, "The Development, Field Test and Validation of an Inventory of Scientific Attitude," *Journal of Reseach in Science Teaching*, 7: 85-94, (1970).

12. Bloom B.S. (Ed.), *Taxonomy of Educational Objectives, Handbook*

I: Cognitive Domain. Ann Arbor, Mich., David Mckay Co., Inc., 1956.

13. Krathwohl, D.R., B.S. Bloom, and B.B. Misia, *Taxonomy of Educational Objectives, Handbook II:Affective Domain*, New York, David Mckay Co., 1964.

14. Eiss, A.F., "The NSTA Conferences on Scientific Literacy," *The Science Teacher*, 35: 30-32, (1968).

15. Eiss, A.F., and M.B. Harbeck, *Behavioral Objectives in the Affective Domain*, Washington, National Science Supervisors Association, 1969.

16. Roakeach, M., *Beliefs, Attitudes and Values*, San Francisco, Jassey-bass Inc., 1968.

17. Engman, B.D., "Behavioral Objectives: Key to Planning," *The Science Teacher*, 15, 86-87, (1968).

18. Thurstone, L.L., "Attitude Can Be Measured," *American Journal of Sociology*, 33, 529-534, 1928; In *Attitude Theory and Measurement*, M. Fishbein, Ed., New York, John Wiley & Sons, Inc., 1967.

19. Thurstone, L.L., "The Measurement of Social Attitude," *The Journal of Abnormal and Social Psychology*, 26: 2149-299, 1931; In *Attitude Theory and Measurement*, M. Fishbein, Ed., New York, John Wiley & Sons, Inc. 1967.

20. Toregerson, W.S. *Theory* and *Methods of Scalling*, New York, John Wiley & sons, Inc., 1958.

21. Likert, R., "The Method of Constructing an Attitude Scale," Archives of Psychology, 1932, pp. 44-53; In *Attitude Theory and Measurement*, M. Fishbein, Ed., New York, John Wiley & Sons, Inc., 1967.

22. Osgod, C.C., G.E. Suci, and P.H. Tannenbaum, *The Measurement of Meaning*. Urbana, Ill., University of Illinois Press, 1957.

23. Oppenheim, A.N., *Questionnaire Design and Attitude Measurement*, New York, Basic Books, Inc., 1956.

24. Shaw, M.E. and J.M. Wright, *Scales for the Measurement of Attitudes*, New York, McGraw-Hill Book Co., 1967.

25. Cronbach, L.J., "Response Sets and Test Validity," *Educational and Psychological Measurement* , 6: 475-494, (1946).

26. Cronbach, L.J., "Further Evidence on Response Sets and Test Designs," *Educational and Psychological Measurement*, 10: 3-31, (1950).

27. Diederich, P.B., "Components of the Scientific attitude," *The Science Teacher*, 334; 23-24, (1967).

28. Obourn, E.S., and L.K. Johnson, *An Analysis and Checklist on the Problem Solving Objective*. U.S. Department of Health, Education, and Welfare, Circular No. 381 (rev.), August, 1960.

29. Hedges, W.D., *Testing and Evaluation for the Sciences in the Secondary School* Belmont, Stare Wadsworth Publishing Co., 1966.

30. Klopfer, L.E., *History of Science Cases*, Chicago, Science Research Associates Inc., 1964.

31. Gulliksen, H., *Theory of Mental Tests*, New York, John Wiley & Sons, Inc., 1950.

32. Loevinger, J., "Objective Tests as Instruments of Psychological Theory." In *Problems in Human Assessment*, D.N. Jackson and S. Messick, Eds., New York, McGraw-Hill, Inc, 1967.

33. Cronbach, L.J., and P. E. Meehl, "Construct Validity in Psychological Tests," *Psychological Bulletin*, 52; 281-302, (1955).

34. Magnisson, D., *Test Theory*. Palo Alto, Calif.,Addison-Wesley Publishing Company, 1966.

35. Harris, C.W., and H.F. Kaiser, "Oblique Factor Analytic Solution by Orthogonal Transformations," *Psychometrika*, 29: 347-362, (1964).

30

The Development and Application of a Scale for Measuring Scientific Attitudes

Victor Y. Billeh
George A. Zakhariades

INTRODUCTION

In the last five decades particular emphasis has been given to the development of scientific attitudes through formal education. Several attempts by many educators were made to define the term "scientific attitude" and to construct appropriate instruments for measuring such attitudes.

One of the first instruments for measuring the scientific attitudes was constructed by Curtis[1] in 1924. The test was based on a definition of the scientific attitude consisting of four components, namely, (a) Conviction of universal cause-and-effect relationships, (b) Habit of delayed response, (c) Habit of Weighing evidence, and (d) Open-mindedness.

In 1935 Noll[2] constructed a new instrument based on another definition consisting of six components of the scientific attitude, namely, (a) Accuracy in all operations including calculation, observation and report, (b) Intellectual honesty, (c) Open-mindedness, (d) The habit of suspended judgement, and (b) the habit of criticism.

A more comprehensive definition was presented in the *Forty-Sixth Yearbook* of the National Society for the study of Education[3].

This detailed list of the characteristics of the scientific attitude is different in content than the definitions, of Curtis and Noll. Another condensed list of habits constituting scientific attitude was derived from the research studies up to 1960 in the *Encylopedia of Educational Research*[4].

Another instrument for measuring the scientific attitudes, the Scientific Attitude Inventory (SAI), was developed and reported by Moore and Sutman[5]. A Likert type scale was constructed and some of the attitudes measured by the scale were:

> The laws and/or theories of science are approximations of truth and are subject to change.
>
> Science is an idea-generating activity and its values in its theoretical aspects.
>
> Progress in science requires public support.

A major criticism of the SAI is that it does not measure the scientific attitude as such, but rather a mixture of attitudes concerning science, sociology of science, and knowledge of the nature of science.

Among the recent presentations of the components of the scientific attitude are those by Haney[6] and Diederich[7]. The definition of Haney consists of eight components and that of Diederich consist of twenty.

Identification of Scientific Attitudes

The definitions of the "scientific attitude" cited above, and especially the last two, were considered carefully for identifying the following six components of the term "scientific attitude."

1. Rationality
   a. Commitment of the value of rationality
   b. Tendency to test traditional beliefs
   c. Seeking for natural causes of events and identification of cause-and-effect relationships.
   d. Acceptance of criticalness
   e. Challenge of authority
2. Curiosity
   a. Desire for understanding new situations that are not

explained by the existing body of knowledge.

b. Seeking to find out the ('whys') and "hows" of observed phenomena

c. Give emphasis on the questioning approach for novel situations

d. Desire for completeness of knowledge.

3. Open-mindedness

a. Willingness to revise opinions and conclusions

b. Desire for new things and ideas

c. Rejection of singular and rigid approach to people, things, and ideas

4. Aversion to Superstitions

a. Rejection of Superstitious beliefs

b. Acceptance of scientific facts and explanations

5. Objectivity-Intellectual Honesty

a. Demonstration of the greatest possible concern for observing and recording facts without any influence of personal pride, bias, or ambition

b. In interpreting results, does not allow any modifications according to present social, economic, or political conditions.

6. Suspended Judgement

a. Unwillingness to draw inferences before evidence is collected.

b. Unwillingness to accept as facts things that are not supported by convincing proof

c. Avoidance of quick judgements and jumps to conclusions.

The above six general components of the scientific attitudes with the specific behavior of each were used for the construction of an instrument for measurement the scientific attitudes.

The Significance of Scientific Attitudes

Emphasis on the development of the scientific attitudes was pointed out by Heiss[8] who stated that the development of the scientific attitudes and the ability to use the methods of science are major goals of science instruction. It was further pointed out that considerable attention has been given to ways of teaching the scientific method, whereas less attention has been given to the development of scientific attitudes. In his suggestions for developing the scientific attitudes. Heiss asserted that the main handicap in this area was the lack of an appropriate instrument for measuring these attitudes.

In explaining the significance of the scientific attitudes, Noll[9] reported that the possession of the scientific attitudes by people is not great importance in helping them understand and properly interpret the scientific knowledge and method as well as many other things concerning our daily lives. This statement is in agreement with the major goal of contemporary science education, namely, the development of scientifically literate citizens[10]. Following Bloom's *Taxonomy of Educational Objectives*, Klopfer[11] established a 48-category system of objectives for science instruction. One of the objectives of the affective domain is the development of scientific attitudes. This provides recognition of the importance of these attitudes to contemporary science education.

THE PROBLEM

The purpose of this study was to: (a) construct a Thurstone-Chave type scale to measure the scientific attitudes of secondary school pupils, university students, and science teachers, (b) compare the scientific attitudes of secondary school pupils, university students and science teachers, and (c) correlate the scientific attitude of secondary school pupils with their achievement in science.

In specific, the following the null hypotheses are investigated: (1) There is no significant difference among the mean scores of tenth graders, twelfth graders, sophomores, seniors, and science teachers on the Scientific Attitude Scale (SAS) constructed in this study; and (2) There is no significant correlation between scores on SAS and science achievement scores for secondary school students in grades ten and twelve.

PROCEDURE

Construction of the Scale

Taking into consideration the characteristics of several types of attitude scales, it was decided to construct the Scientific Attitude Scale in the Thurstone-Chave form.

A total number of 87 items concerning the scientific attitude, as they were identified earlier in this study, were originally constructed and presented to a panel of judges for determining the scale value (S) and the measure of variation (Q) for each statement. For this purpose a sample of 45 scientists was selected to form the group of judges. Thirty of the scientists were professors of Biology, Chemistry, Physics and Agriculture Sciences at the American University of Beriut. The Other scientists were selected from the staff of the Institute of Agricultural Research and other scientists in Cyprus. According to Oppenheim[12] 45 Judges are considered a sufficient group for the construction of the present scale.

The judges were asked to sort each statement in one of the eleven piles, where the pile No. 11 indicates the most favorable feelings toward the psychological object, pile No., 1 indicates the most unfavourable feelings, and pile No. 6 is determined as a neutral point expressing neither favorable nor unfavorable feelings.

The scale value, S, of each statement was obtained by the median of the distribution of judgement for each statement. The Q value was found by calculating the interquartile range of the distribution of judgement for each statement.

On the basis of the S and Q values, 46 statements were selected. The scale values of the selected statement covered the range of the continuum and had low Q values, Statements with Q values above 3.0 were eliminated.

The Pilot Study

The constructed scale was administered to two classes of tenth graders and to two classes of twelfth graders in a public secondary school in Famagusta. The subjects participated in this pilot study were 88 and they were asked to respond to each statement with "Agree" or "Disagree".

From the results of this testing the indices of difficulty and discrimination were calculated from each statement. Out of the 46 statements ten were eliminated because in low discrimination indices. Some of the items were revised in terms of language. Thus the final form of the scale was developed (see Appendix).

The Sample

Out of the total number of 47 Greek public secondary schools in Cyprus, three were randomly selected. From each school, two classes of tenth graders and two classes of twelfth graders were selected. The science teacher working in the three selected schools and an additional group working in other schools constituted the group of science teachers. The group of sophomores consisted of Chemistry and Biology majors and groups of seniors consisted of students majoring in chemistry, biology, and physics at the American University of Beirut. So, A sample of 349 secondary school pupils, 121 university students, and 31 science teachers was formed.

The final scale was administered to all the subjects. The achievement in science for the secondary school pupils was obtained from the school records.

Analysis of Data

The reliability of the scale was calculated by the split-half technique and the Spearman-Brown formula was applied for the correlation. To test the first hypothesis the one-way analysis of variance was employed. Newmann-Keuls Test[13] was then utilized to determine wherein significant difference occurred.

To test the second null hypothesis, the Pearson correlation coefficient between attitude scores and science achievement scores was calculated and tested for significance.

RESULTS

In the light of the results of the items analysis a final scale of 36 items was constructed. The Statements were classified in such a way so that the scale was built in two equivalent halves, Half A and Half B. Each statement in Half A had a corresponding statement in Half B with approximately the same S and Q values. The structures of this scale is shown at Table 1. From this structure of the scale the following formulas were derived for obtaining the score of each individual on the scale.

Where the numbers 83.5 and 84.7 are the sums of the scale values of all the positive statements in Half A and Half B, respectively.

The numbers 18.0 and 17.4 are the sums of the scale values

Table 1
Structure of the Final Scale

Component	No. of Statements				S Values			
	Half A		*Half B*		*Half A*		*Half B*	
	Pos	Neg	Pos	Neg	Pos	Neg	Pos	Neg
Rationality	1	3	0	3	10.3	5.2	0.0	6.6
Curiosity	1	1	2	1	7.0	2.7	19.3	2.5
Open-Mindedness	2	1	1	2	20.2	1.5	8.6	3.0
Aversion to Superstitions	1	2	1	1	9.9	2.6	8.5	1.4
Object-Intell. Honesty	3	0	2	1	27.3	0.0	20.5	2.7
Suspended Judgement	1	2	3	1	8.8	6.0	27.8	1.2
Total	9	9	9	9	83.5	18.0	84.7	17.4
	18		18		101.5		102.1	

$$\text{Score on Half A} = \frac{99}{18}\left(\frac{\Sigma X_A}{83.5} + \frac{\Sigma Y_A}{18.0}\right)$$

$$\text{Score on Half B} = \frac{99}{18}\left(\frac{\Sigma X_B}{84.7} + \frac{\Sigma Y_B}{17.4}\right)$$

$$\text{Score on the Scale} = \frac{A+B}{2}$$

of all the negative statements in Half A and Half B, respectively.

ΣX_A and ΣX_B represent the sum of the scale values for the correct responses to the positive statements in Half A and Half B, respectively.

ΣX_A and ΣX_B represent the sum of the scale values for the correct responses to the positive statements in Half A and Half B, respectively.

ΣX_A and ΣX_B represent the sum of the scale values for the correct responses to the negative statements in Half A and Half B, respectively.

For obtaining the sub scores on the six components of the scale similar formulas were derived.

The Pearson's correlation coefficient was calculated for scores for Half A and Half B. The Spearman-Brown formula was applied for the correlation and the reliability of the scale was thus determined for each group. The reliability coefficient ranged from 0.55 to 0.74. The reliability of the Science Scale constructed by Brandyberry for measuring the scientific attitude ranged between 0.50 and 0.78 and it was considered be relatively[14]. In another study conducted by Weinhold[15], the instrument used for measuring the scientific attitudes of elementary school teachers had a reliable of 0.70. Comparing the reliability of the constructed scale to the reliabilities of other similar instruments it can be concluded that the scale has a satisfactory and acceptable reliability.

The obtained means and standard deviations are presented in Table II.

Table II
Means and Standard Deviations of Attitude Scores

Group	Mean	St. Deviation
Group A, 10th graders	8.11	1.08
Group B, Sophomores	8.59	1.07
Group C, 12th graders	8.89	1.03
Group D, Seniors	9.28	0.91
Group E, Science teachers	9.35	0.76

It can be observed that all groups have mean scores above 6.0 and so the scientific attitudes are positive in all cases. It is also clear that the mean score of the tenth graders is the lowest and that of the science teachers is the highest.

To determine whether significant difference exist among the mean scores of the five groups the one-way analysis of variance was employed. A summary of the results is presented in Table III.

Table III
One-Way Analysis of Variance for Differences Among Mean Scores

Source	S.S.	df	M.S.	F
Between groups	98.688	4	24.692	23.486**
Within groups	521.036	496	1.050	
Total	619.724	500		

**p<0.01

The obtained value of F ratio indicates that there are significant difference among the five mean scores. To determine which groups differ significantly in their mean scores the Newman-Keuls multiple comparison test was employed. The results of this test are summarized in Table IV.

It can be seen that significant differences exist among the mean scores of the science teachers, tenth graders, twelfth graders, and university sophomores. The mean score of university seniors was significantly higher than those of the tenth graders, twelfth graders and university sophomores. The twelfth graders and sophomore students scored significantly higher than the tenth graders. No significant differences were revealed between the mean scores of sophomore students and twelfth graders and between those of science teachers and the senior students. The science teachers are holders of Bachelor of Science degrees in physics, chemistry, or biology, and the university senior were about to receive their Bachelor of Science degrees when the scale was administered to them.

Table IV
Newman-Keuls Test for Differences Among Man Scores

Group	A	B	C	D	E
A	--	0.48**	0.78**	1.17**	1.24**
B		--	0.30	0.69**	0.76**
C			--	0.39*	0.66**
D				--	0.07
E					--

**p<0.01
*p<0.05

The Pearson's r correlation coefficient between the scientific attitude scores and the grades in science was calculated for the twelfth graders. The obtained values for the two groups were 0.247 and 0.248, respectively. Both values are significantly different from zero at the 0.02 level.

CONCLUSIONS

The results presented above point to several observation which can be made concerning the scientific attitudes of secondary school pupils, university students, and science teachers.

1. All the groups of subjects: secondary school pupils, university students, and science teachers demonstrated positive scientific attitude.
2. The obtained difference among the mean scores on SAS of the five groups reveal that the amount of science knowledge or the general exposure of science courses effects positively the scientific attitude.
3. It seems that the courses the students had from twelfth grade to sophomore did not influence the scientific attitude. On the other hand, the science courses at the university junior and senior levels appeared to be effective in developing positive changes on the scientific attitudes of the students.
4. Since no significant difference was found between university seniors and science teachers one may conclude that experience in teaching secondary school science does not seem to be effective in developing further the scientific attitudes.
5. A low positive relationship exists between the scores on the SAS and the grades received by secondary school students in science.

The results of the study concerning the constructed scale may lend to the following conclusions:

1. The fact that the amount of knowledge in science or exposure to science seems to be a variable influencing the scientific attitudes, as measured by SAS, may be considered as an indication for the construct validity of the instrument. This can be supported by the bindings of Seymour and Sutman[16] and the ideas of Dressel.[17]
2. The identification of the six general components of the scientific attitude and the construction of the statements according to the behavior of each component may be considered an appropriate procedure for assuming the content validity of the constructed scale.

APPENDIX

The Scientific Attitude Scale (SAS)

Item No.		*S*	*Q*
1.	Knowledge is promoted if every new idea in a field is accepted immediately after it is reported.	1.6	1.8
2.	Whenever there is insufficient evidence either for or against a proposition, the wisest thing is neither to accept it nor to reject it.	9.9	1.7
3.	Listening to new ideas is a very interesting and pleasing activity.	9.7	2.4
4.	If one of our ideas is poor, we should hold this idea although it is proved to be poor.	1.8	1.8
5.	Students who are keen to learn new ideas despite their inadequate knowledge are likely to became scientists.	7.0	2.4
6.	The records of the observations of a scientist reflect the personal feelings of the scientist.	2.7	2.2
7.	In unscientific discussions one may hear that someone is willing to prove that a certain idea or opinion is absolutely correct.	8.1	3.0
8.	Experimental testing of new idea is mainly carried out to satisfy the person who first suggest the idea.	3.0	3.0
9.	Fortune tellers usually flourish in scientific communities.	1.3	1.9
10.	Knowledge, once accepted, is not subject to change.	1.2	1.2
11.	Scientific explanation should be preferred to the romantic stories of astrologers and magicians.	9.9	2.3

Item No.		S	Q
12.	Dr. E., a very famous scientist, presented a new theory and Dr. A., a young scientist who had graduated two years before, had some doubts about this theory. Dr. A. should accept the theory because it was presented by Dr. E.	1.4	1.4
13.	People should be willing to change their ideas if enough evidence shows that the ideas are poor.	10.5	1.7
14.	In interpreting the results of his work, a scientist should not be affected by external social conditions.	9.9	2.1
15.	If a new facts agree with some ideas it can be concluded that these ideas are true.	2.7	2.2
16.	A questioning attitude should dominate the approach to any novel situation.	9.5	2.3
17.	Explanations can only be made if observable data can be collected.	9.4	1.9
18.	One should not criticize the work of others.	2.2	2.4
19.	Novel situations which cannot be explained with the existing body of knowledge are undesirable.	2.7	2.9
20.	Questioning opinions and ideas has nothing to do with scientific behavior.	2.5	2.6
21.	One should express one's judgement about new ideas and new discoveries immediately after one learns of them.	3.3	2.5
22.	Newly discovered ideas should be reported unchanged even if they contradict existing ones.	10.6	1.3
23.	The editor of a well-known journal should not accept for publication research studies of beginners.	1.5	1.7

Item No.		S	Q
24.	Meteorologists can predict the weather with high probability.	8.5	1.9
25.	When traditional beliefs are in conflict with scientist discoveries, it is better to accept the traditional beliefs.	2.4	1.7
26.	It is worthless to listen to a new idea unless all people accept it.	1.2	0.8
27.	A successful scientist is more objective than a politician.	9.8	2.2
28.	Knowledge in a country is promoted if local publication include tne writings of famous and unknown writers.	8.6	1.4
29.	The largest attendance at cinemas occurs on Saturday evenings and the largest attendance at church occur on Sunday mornings. Whenever there are many people watching a film, the following day many people should go to church.	1.2	1.2
30.	Astrology is a science which contributes to the better understanding of people.	1.4	1.5
31.	Knowledge should be considered tentative.	8.8	2.5
32.	Seeking further information about a novel situation is considered a sound approach to begin with.	9.8	2.3
33.	Criticism benefits the advancement of knowledge.	10.3	1.1
34.	Intelligence is the main factor that contributes to the advancement of knowledge.	7.7	2.6
35.	For weather predictions, magicians, and astrologers should be consulted.	1.3	1.3
36.	Enough evidence supporting a certain idea should be provided before the idea is accepted.	10.2	2.0

REFERENCES

1. Noll, V.H., *The Teaching of Science in Elementary and Secondary Schools*, New York: Longmans Green and Co., 1942, p. 228.
2. Ibid., p, 25.
3. National Society for the Study of Education, "Science in the Secondary School," In *Science Education in American Schools*, The Forty-Sixth Yearbooks Part I, Chicago: The University of Chicago Press, 1947, pp. 147-149.
4. Harris, C. (Ed.), "Science: Aims and Objectives," in *The Encyclopedia of Educational Research*, 3, 1960, p. 1221.
5. Moore, R.W. and Sutman, F.X. " The Development, Field Test and Validation of an Inventory of Scientific Attitudes," *Journal of Research in Science Teaching*, 6, 85-94, 1970.
6. Hancy, R.E., "The Development of Scientific Attitudes," *The Science Teaching* 31: 33-35, 1964.
7. Diedderich, P.B., "Components of the Scientific Attitude," *The Science Teacher*, 34: 23-24, 1967.
8. Heiss, E.D., "Helping Students Develop a Scientific Attitude," *The Science Teacher*, 25:. 371-373, 1958.
9. Ref. I, p.23.
10. NSTA Position Statement, " School Science Education for the 70's, *The Science Teacher* 38: 46-51, 1971.
11. Bloom, B.S. Hastings, I., and Madaus, G., *Handbook on Formative and Sammative Evaluation of Student Learning*, New York: Harper and Row, Publishers, 1971, pp, 559-641.
12. Oppenheim, N.A., *Questionnaire Design and Attitude Measurement*, New York: Basic Books, Inc., Punlishser, 1966, p.126.
13. Winer, B.J., *Statistical Principles in Experimental Design*, New York: Mc- Graw Hill Book Compnay, 1962, pp. 101-104.
14. Murphy, G., "Content Versus Process Centeral Biology Laboratories, Part II: The Development of Knowledge, Scientific Attitude, Problem- Solving Ability and Interest in Biology," *Science Education*, 52; 148-162, 1968.
15. Weinhold, J.d., " An Attempt to Measure the Scientific Attitude of Elementary School Teachers," *Dissertation Abstracts International*, 31; 1647-A, 1970.
16. Seymour, L. and Sutman,F.X., "Critical-Thinking Ability, Open-Mindedness, and Knowledge and of the Processes of Science of

Chemistry and Non-Chemistry Students." *Journal of Research in Science Teaching*, 10: 159-163, 1973.

17. Dressel, P.L. et. al., "How the Individual learn Science," *Rethinking Science Eduction*. The Fifth-Ninth Yearbook of the National Society for the study of Education, Part I. Chicago: The University of Chicago Press, 1960, p.47.

31

Scientific Attitude Scale

N.N. Shrivastava

INTRODUCTION

The scientific attitude may be defined as "Open-Mindedness", a desire for accurate knowledge, confidence in procedures for seeking knowledge and the expectation that the solution of the problem will come through the use of varified knowledge.

The scientific attitude recognises three cardinal principles, firstly, truth is what it is searching for, secondly, no ways should be missed that might help to find truth and thirdly, what may seem to be the truth at one time, may later, under the advance of new facts prove to be something less than the truth. For this reason a man with scientific attitude withholds judgement which means that he simply keeps his mind and eyes open for anything new that may prove better than the old. Thus the true scientific or naturalist would not take the apparent facts shown in the experiment as the final answer. He would not be satisfied with such answer until the experiment had carried on for many months, perhaps for years.

The major components of the scientific attitude are as below—

1. Rationality

1. Commitment of the value of rationality.
2. Tendency to test traditional beliefs.
3. Seeking of natural cause of events and identification of cause and effect relationship.

4. Acceptance of criticalness.
5. Challenge of authority.

2. Curiosity

1. Desire for understanding new situations that are not explained.
2. Findout the "whys" and "hows" of observed phenomenon.
3. Give emphasis on the questioning approach of noval situations.
4. Desire for completeness of knowledge.

3. Open-mindedness

1. Willingness to revise opinions and conclusions.
2. Desire for new things and ideas.
3. Rejection of singular and original approach to people, things and ideas.

4. Aversion to Superstitions

1. Rejection of superstitious belief.
2. Acceptance of scientific facts and explanations.

5. Objectivity—Intellectual Honesty

1. Demonstration of the greater possible concern for observing and recognizing facts without any influence of personal pride, bias or ambition.
2. In interpreting results, does not allow any modifications according to present social, economic or political situations.

6. Suspended Judgements

1. Unwillingness to draw inference before evidence is collected.
2. Unwillingness to accept as facts, things that are not supported by convincing proof.
3. Avoidance of quick judgements and jumps to conclusions.

Objective

Scientific attitude is the most important outcome of the science teaching. The students should be made aware about the facts and they should develop the tendency of varifying the facts in a scientific manner. Development of scientific attitude is one of our major objective of teaching science in our educational institutions. Hence the scale will enable us the effectiveness of teaching science with an objective of developing scientific attitude.

Construction of SAS

In the scientific attitude scale the statements of all the six components were included. Originally 90 statements were constructed on the basis of Likert method. Finally only 36 statements were retained. The scale value of the adopted scale was varified by the method of Thurstone equal appearing interval for selection of items.

Reliability

Method	No.	Coefficient of Correlation	
		Teachers	*Students*
Split-half Method	100	0.82	0.86
Equivalent Form Method	100	0.90	0.88

Validity

	No.	Coefficient of Correlation	
		Teachers	*Students*
Index of reliability	100	0.94	0.90

SAMPLE

The study was confined to Science and non-science teachers and science and non-science students of the educational division headquarters of the State of Madhya Pradesh. A random sample of 50 science and 50 non-science male and female teachers and 100 science and 100 non-science male and female students was drawn from different schools located at division headquarters. The age range of the students was 14 to 22 and the teachers age range was 25 to 55 years.

Instructions

There are few statements in this scale. Every statement has three alternative options 'Agree', 'Undecided' and 'Disagree'. If you agree with the statement put a right (✓) mark in the box against it, if you are not in a position to decide put the right mark in the box indicating 'Undecided' and if you do not agree with the statement, put the right mark in the box against 'disagree'.

There is no time limit to complete the scale but you try to complete it as quickly as you can."

Scoring

Allotment of marks is as below :

2 points for agree
1 point for undecided
0 point for disagree

Statistical Values

	Mean	*S.D.*	*Median*
Teachers	47.80	7.82	54.37
Students	38.75	6.89	42.28
Combined group	44.25	7.55	45.25

Correlation Matrix

	Rationallity	*Curiosity*	*Open-mindedness*	*Aversion to superstitions*	*Objectivity*	*Susp. Judgement*
	1	*2*	*3*	*4*	*5*	*6*
1. Rationality	--	.42	.29	.31	.62	.43
2. Curiosity	.42	--	.43	.51	.83	.83
3. Open Mindedness	.29	.43	--	.71	.41	.32
4. Aversion to superstition	.31	.51	.72	--	.66	.88
5. Objectivity Intellectual Honesty	.62	.83	.41	.65	--	.41
6. Suspended Judgement	.42	.83	.32	.38	.41	--

Conversion of raw scores into T scores

Raw Scores	T Scores Teachers	T Scores Students	Raw Scores	T Scores Teachers	T Scores Students
28	25	35	52	55	69
32	30	41	56	60	75
36	35	47	60	65	80
40	41	52	64	70	86
44	46	57	68	75	92
48	50	63	72	80	98

Percentile norm for Students and Teachers

Percentiles	Students	Teachers
P_{90}	51.21	59.68
P_{80}	47.68	56.42
P_{70}	44.54	53.61
P_{60}	41.36	50.26
P_{50}	38.64	47.34
P_{40}	35.21	45.52
P_{30}	32.75	42.61
P_{20}	29.54	38.52
P_{10}	26.55	35.42

Interpretation of raw scores :

Classification of Scientific attitude	Teachers	Students
Very High	60 & above	51 & above
High	52–59	43–50
Moderate	44–51	35–42
Low	36–43	27–34
Very Low	35 & below	26 & below

SCIENTIFIC ATTITUDE SCALE

Name..

Age............................Sex............................Class........................

School/College..

Address..

....

....

Instructions

There are a few statements in this Scale. Every statement has three alternative options, viz., Agree, Undecided and Disagree. If you agree with the statement you put a right mark (✓) in the box against to it, if you are not in a position to decide you put the right mark in the box against to it, and if you do not agree with the statement you put the right mark in the box against to it.

There is no time limit to complete the Scale, but you try to complete it as quickly as you can.

Scores Obtained	Percentage of Scores	Classification of Scientific Attitude

S. No.	Statement	Agree	Un-decided	Dis-agree
1.	Immediate publication of newly derived facts helps in the development of knowledge.	☐	☐	☐
2.	Any hypothesis should not be favoured or disfavoured in the absence of sufficient proof.	☐	☐	☐
3.	Knowing of new ideas will be interesting and delighting.	☐	☐	☐
4.	A view should not be changed even it is proved insignificant.	☐	☐	☐
5.	A curious person, though immature, can become a scientist.	☐	☐	☐
6.	A scientist's record is the reflection of his views and personality.	☐	☐	☐
7.	Some people, often, try to express any of their views in the general discussions.	☐	☐	☐
8.	People want to satisfy themselves by proving their new ideas through experimentation.	☐	☐	☐
9.	Among scientists astrologers are more efficacious.	☐	☐	☐
10.	Any accepted knowledge is unchanging.	☐	☐	☐

Contd.

S. No.	Statement	Agree	Undecided	Disagree
11.	Thrilling stories of the astrologers and magicians should be given importance without any scientific reasoning.	☐	☐	☐
12.	A theory proposed by a famous should be accepted.	☐	☐	☐
13.	If any current concept is proved false by several proofs, the people must be ready to change the concept.	☐	☐	☐
14.	A scientist should not be influenced by any external social forces in revealing his results clearly.	☐	☐	☐
15.	Some positive facts of a concept conclude that the whole concept is true.	☐	☐	☐
16.	Logical nature helps before assuming any situation.	☐	☐	☐
17.	Results can be derived only after assimilating check figures.	☐	☐	☐
18.	Any new thing cannot be criticised in the absence of facts.	☐	☐	☐
19.	Known is the basis to know unknown.	☐	☐	☐
20.	There is no relation between questioning nature and scientific approach.	☐	☐	☐

Contd.

S. No.	Statement	Agree	Undecided	Disagree
21.	A person should express his decisions only on the basis of new inventions and achievements.	☐	☐	☐
22.	Inventions should be revealed even they are against to the current concpets.	☐	☐	☐
23.	Theses of the new researchers too should be published in important journals.	☐	☐	☐
24.	Metereologists only can presume weather.	☐	☐	☐
25.	Traditional concepts should be accepted only when they differ scientific research.	☐	☐	☐
26.	It is useless to hear/observe fresh views until they are accepted by all.	☐	☐	☐
27.	A successful scientist is more materialistic than a politician.	☐	☐	☐
28.	Knowledge is improved in any country only when the articles of unnamed writers are published in local journals.	☐	☐	☐
29.	Crowds at religious places on Saturdays and at cinema theatres on Sundays say that cinemas should be watched after visiting religious places.	☐	☐	☐

Contd.

S. No.	Statement	Agree	Un-decided	Dis-agree
30.	Astrology is the science of knowing people.	☐	☐	☐
31.	Testing of knowledge must be procedural.	☐	☐	☐
32.	Concerned information should be possessed to know its modern position.	☐	☐	☐
33.	Criticism is useful in the acquisition of knowledge.	☐	☐	☐
34.	Development of knowledge is having relationship only with the mind.	☐	☐	☐
35.	We should not depend on the presumptions of astrologers and magicians regarding the weather.	☐	☐	☐
36.	Sufficient proofs should be collected before accepting an idea.	☐	☐	☐

32

Scientific Attitude Scale

J. K. Sood
&
R. P. Sandhya

CONSTRUCTION AND VALIDATION OF SCIENTIFIC ATTITUDE SCALE (SAS)

The survey of related literature has revealed that during the past five decades a number of instruments have been developed for measuring, scientific attitude. Haff, A.Q. (1936), Noll (1935) Moore and Sutman (1970) Kozlow and Nay (1975) have developed such instruments. Most of the instruments suffer from one or the other short-comings. There is tendency to lump together several dimensions of science under the caption of attitude such as ; interest, attitude and values. The content covered in these instruments do not suit the Indian conditions, because of such short-comings. Therefore a new tool was developed for measuring the scientific attitude of Indian teachers. Some scalers have been developed by Indian researchers, too.

Theoretical Frame work

In 1975 Billech and Zakharidas had developed an instrument which comprises of six dimensions, namely; Rationality, Open-mindedness; Curiosity, Aversion to Superstitions; Objectivity of Intellectual beliefs, and Suspended judgement.

The Investigator has taken Billech and Zakharidas's theoretical assumptions as bases to develop his own tool concerning scientific attitude.

These six dimensions reflects the followings views in brief:

Dimensions I Rationality:

a) Commitment of the value of rationality.

b) Tendency to test traditional beliefs.

c) Seeking for natural cause of events and identification of cause-and-effect relationships.

d) Acceptance of criticalness.

e) Challenge of authority.

Dimensions II-Curiosity:

a) Desire for understanding new situations that are not explained by the existing body of knowledge.

b) Seeking to find out the 'why' and 'Hows' of observed phenomena.

c) Give emphasis on the questioning approach for novel situation.

d) Desire for completeness of knowledge.

Dimension III-Open-mindedness:

a) Willingness to revise opinions and conclusions.

b) Desire for new things and ideas.

c) Rejection of singular and rigid approach to people; things; and ideas.

Dimensions IV-Aversion to Superstitions:

a) Rejection of superstitions and false beliefs.

b) Acceptance of scientific facts and explanation.

Dimension V-Objectivity of Intellectual Beliefs:

a) Demonstration of the greatest possible concern for observing and recording facts without any influence of personal pride, bias or ambition.

b) Not allowing any change in interpreting results on the basis of present social, economic or political influences.

Dimension VI- Suspended Judgement:

a) Unwillingness to draw inferences before evidence is

collected.

b) Unwillingness to accept facts that are not supported by the convincing proof.

c) Avoidance of quick judgements.

The investigator selected the Likert Method to construct Scientific Attitude scale due to under-mentioned reasons:

1) It has been claimed by Likert (1932) that the method of summated ratings; in his survey of the attitudes of employed of its unemployed men was adopted because of its relative simplicity. Rundquist and Sletto (1932) used this method in developing the attitude scale contained in the Minnesota survey of opinions and they also expressed their belief that the method was less laborious than that developed by Thurstone.

2) Secondly, it is less time consuming. Edwards and Kenney (1946) in their comparative study of the method of equal appearing intervals and the method of summated ratings, estimated that the time required to construct equal appearing interval scale, was approximately twice that required by the method of summated ratings.

3) Scale constructed by the Likert method yield higher reliability coefficient with fewer items than scales constructed by the Thurstone method. This was the finding arrived at by Hall (1934) in his survey of the attitudes of employed and unemployed men.

Item writing, Editing and Revision.

The investigator reviewed books; periodicals and other descriptive material dealing with the contributions of scientific attitude. These source were found to be very useful in obtaining statements. Most of the statements used in this scale were collected from the literature and contributions of J. J. Schawab; J.S. Burner; Bertant Russel; William D. Romey, P.L. Gordener, Victor Y. Billech, Paul B. Diederich, Curtis, Moll, Haney E. Richard and many others.

The attitude statements were words in accordance with the following suggestions by Edwards and Killpatrick (1948); Wong (1932), Thurstone and Chave (1929), Likert (1932) and Bird

(1940) :

1. Avoid statements that refer to the past rather than to the present.
2. Avoid statements that are factual or capable of being interpreted as factual.
3. Avoid statements that may be interpreted in more than one way.
4. Avoid statements that are irrelevant to the psychological object under construction.
5. Avoid statements that are likely to be endorsed by almost everyone or by almost no one.
6. Select statements that are believed to cover the entire range of the affective scale of interest.
7. Keep the language of statements simple, clear and direct.
8. Statements should be short, rarely exceeding 20 words.
9. Each statement should contain only one complete thought.
10. Statement containing universals such as more and never often introduce ambiguity and should be avoided.
11. Words such as only, just, merely and others of a similar nature should be used with care.
12. Whenever possible, statements should be in the form of simple sentences rather than in form of compound or complex sentences.
13. Avoid the use of words that may not be understood by those who are to be given the completed scale.
14. Avoid the use of double negatives.

An initial pool of 130 statements was prepared. This pool of statements was given to ten experienced and qualified educators after getting, its language checked by experts and ascertaining that vocabulary used was appropriate for teachers. They were requested to rate each statement on three categories by answering the undermentioned question:

Is the attitude/view measured by this item -

-essential

-useful but not essential or

-not necessary

After collecting their opinion on every statement content validity rations (C.V.R.) were calculated by using the following formula (Lawshe 1975):

$$C.V.R. = \frac{ne - \frac{N}{2}}{N/2}$$

where ne = number of penalists indicating an item essential.

N = Number of penalists.

Considering table (See appendix) statements whose CVRs were more than or equal to 0.62 were selected because C.V.R. 0.62, is significant at 0.05 level of significance for N=10. In this way content validity of the statements was ascertained quantitatively by utilizing Lawshe's suggestion. Thus, Out of 130 items only 66 statements were retained.

Initial tryout

Sixty six statements on different dimensions of Scientific Attitude Scale were arranged in Likert Type Instrument.

Fifty percent items were of positive polarity and remaining fifty percent items were of negative polarity (See appendix I) This instrument was administered on science teacher of both sexes. The teachers were asked to assign any one of the five following categories after carefully reading the statement. The five categories were :

SA — Strongly agree,

A — Agree

N — Neutral

D — Disagree

SD — Strongly Disagree

After the administration of instrument, it was scored by keeping into consideration the scoring procedure suggested by Likert.

For SA response 5

For A response 4

For N response	3
For D response	2
For SD response	1

For items of negative polarity, the scoring system was reverse. For finally selecting the item undermentioned procedure suggested by Edwards (1957) was adopted :

Step I

The investigator considered the frequency distribution of scores based upon the response to all statements. Then 25% of the subjects (NH 50) with highest total scores and also 25% of the subjects (NL 50) with the lowest total scores were selected. These were termed as high and low groups.

Step II

In evaluating the responses of the high and the low groups on each statement *t* values were computed by using undermentioned formula:

$$t = \frac{X_H - X_L}{\frac{(X_H - X_H)^2 \; (X - X_L)^2}{n(n-1)}}$$

where X = the mean score on a given statement for the high group.

X_L = the mean score on the same statement for the low group.

n = $n_H = n_L$ (In our case $n = n_H = n_L = 50$)

The 't' value for 66 statement as calculated by using above formula are given in the following table :

Table No. 32.1
't' Values for Different Items

Statement No.	X_H	X_L	't' value
1.	4.08	4.16	–3.875
2.	4.60	4.18	1.600

Contd...

Table No. 32.1 (Contd.)

Statement No.	X_H	X_L	't' value
3.	3.56	2.56	3.090
4.	4.76	4.32	2.96
5.	4.64	3.22	6.873
6.	4.52	3.82	3.043
7.	3.28	2.76	1.796
8.	4.18	3.48	2.973
9.	3.16	2.54	2.743
10.	4.42	3.60	3.703
11.	4.58	3.676	3.87
12.	4.80	3.98	4.25
13.	2.42	2.62	–.644
14.	4.96	4.40	2.75
15.	4.96	4.08	5.462
16.	3.78	3.3	1.795
17.	4.98	4.56	3.067
18.	4.04	4.10	–.242
19.	4.56	3.32	4.732
20.	5.00	4.38	3.906
21.	4.86	4.12	4.465
22.	4.14	2.78	5.03
23.	4.14	4.14	0.00
24.	3.82	2.92	3.024
25.	4.98	4.16	5.086
26.	3.46	2.64	2.408
27.	4.86	2.82	4.634
28.	4.72	3.2	6.312
29.	4.98	3.4	7.563
30.	4.76	3.9	4.308
31.	4.08	3.04	3.977
32.	3.56	2.68	2.598
33.	4.92	4.02	4.074

Contd...

Table No. 32.1 (Contd.)

Statement No.	X_H	X_L	't' value
34.	4.72	3.8	4.121
35.	4.92	3.38	6.701
36.	4.68	4.06	2.930
37.	2.88	3.08	0.600
38.	4.88	3.48	6.594
39.	4.72	3.04	7.352
40.	4.86	3.72	5.068
41.	4.98	3.28	7.296
42.	4.72	3.24	6.040
43.	4.26	3.04	4.610
44.	4.30	3.56	2.853
45.	4.98	3.56	6.787
46.	5.0	3.80	5.251
47.	4.90	3.5	5.922
48.	4.88	3.68	4.948
49.	4.72	3.5	4.736
50.	4.38	3.56	3.168
51.	4.24	3.42	2.863
52.	3.88	3.0	3.13
53.	2.82	2.58	0.08
54.	4.98	3.98	6.097
55.	4.18	2.66	6.301
56.	4.5	3.36	4.639
57.	4.76	2.98	7.497
58.	3.84	2.42	4.605
59.	4.38	3.60	3.416
60.	4.84	3.64	5.65
61.	4.90	4.12	4.37
62.	4.60	3.28	5.333
63.	4.12	3.08	6.29
64.	3.96	3.08	3.636
65.	4.72	3.44	6.26
66.	4.82	3.78	4.15

Step III

From the table 32.1 it is evidence that item No. 1, 13, 18 and 37 had negative discrimination value (because $X_L X_H$) and there by they were rejected at the first sight. In each dimension only six statements, having higher t value, were selected.

Final form of the scale

The final form of the scale contained 36 statements out of which 18 were of positive polarity and 18 were of negative polarity. The distribution of items as achieved after item analysis on different dimension was as follows :

	Dimension	Positive Polarity (Item Nos.)	Negative Polarity (Item Nos.)
I	Rationality	1, 2, 6	3, 4, 5
II	Curiosity	8, 9	7, 10, 11, 12
III	Open-mindedness	13, 14	15, 16, 17, 18
IV	Aversion to superstitions	19, 21, 24	20, 22, 23
V	Objectivity of Intellectual beliefs	25, 26, 28, 30	27, 29
VI	Suspended judgement	31, 32, 34, 35	33, 36

The according procedure for the items of positive polarity is as follows :

for SA response	5
for A response	4
for N response	3
for D response	2
for SD response	1

For the items of negative polarity the scoring procedure is as follows :

for SA response	1
for A response	2
for N response	3
for D response	4
for SD response	5

The maximum and the minimum score, which the teachers may score of SAS may be 180 36 respectively.

Internal consistency and discriminant validity

The reliability of Scientific Attitude Scale (SAS)' score as calculated by Split-half-method was found to be 0.88. Scientific Attitude Scale was administered on 200 teachers. Data from the administration were analyzed to provide information about the internal consistency and discriminant validity of the dimensions. The results are summarized in Table 32.2 as follows :

Table 32.2
Internal Consistency (Reliability Co-Efficients to and Discriminant Validity Scale Inter-Correlations) of Different Dimensions of Scientific Attitude Scale

Dimension/ Sub scale	No. of Ite-ms	Re-lia-bili-ty	Internal Correlation Coefficient					
			I	II	III	IV	V	VI
1. Rationality	6	.76	1.0	.35	.31	.34	.33	.32
2. Curiosity	6	.86	3.5	1.0	5.3	4.8	.50	.50
3. Open-mindedness	6	.84	3.1	.53	1.0	.53	.49	.53
4. Aversion to super-stition	6	.80	.34	.48	.53	1.0	.57	.53
5. Objectivity of intellectual beliefs	6	.73	.33	.50	.49	.57	1.0	.63
6. Suspended judge-ment	6	.82	.32	.50	.53	.53	.63	1.0
7. Scientific attitude scale (complete)	36	.86	.62	.76	.80	.81	.85	.82

Reliability coefficient for six dimensions of Scientific Attitude Scale were found to be. 76, .86, .84, .80, .73 and .82 respectively. These high coefficients reflect the strength of each dimension for measuring, the Scientific Attitude. In addition to this the high correlation-coefficient between the total attitude scores and the scores on various dimensions taken separately justify inclusion of these dimensions in the attitude scale. From table 3.3 it is evidence that the correlation coefficients was found to be .62, .76, .80, .81, .82 respectively.

APPENDIX

SCIENTIFIC ATTITUDE SCALE

Directions

The following statements are concerned with Scientific attitude.

Read each statement carefully and then mark your answer on the answer sheet provided. Work rapidly.

Record your first impression, the feeling that comes to your mind, as you read the item.

Draw a circle around AA if you fully agree with an item.

Draw a circle around A if you are in partial agreement.

Draw a circle around N if you are neutral.

Draw a circle around D if you partial disagree.

Draw a circle around DD if you totally disagree.

Example—Tajmahal is a beautiful building.

(AA) A N D DD

Since AA is circle this, it means that you completely agree with the statement concerned.

Answer all statements, you have 36 minutes which should give sufficient to finish all 36 items.

Please do not mark on the test booklet.

1. Now it is not possible to develop more sensitive X-ray machine.
2. To challenge the Bible that "the sun revolves round the earth" was not the right step of Copernicus.

3. A conclusion based on insufficient evidences should neither be accepted nor be rejected.
4. An idea should not be accepted if it is proved to be poor.
5. The Scientists should have to find out the occurrences of the undesired events in nature.
6. The ideas are true if few facts support them.
7. Scientists should be curious to find out the cause of the abnormal birth (like two heads and eight legs) of a calf.
8. Every novel situation should not be viewed in an interrogative way.
9. After observing the astonishing situations, like magic, a person should not strive to know the secret of it.
10. Until they achieve success, the scientist should continue their efforts in collecting complete information about the Mars.
11. Scientists shall be able to forecast the sex of a foetus in future.
12. Science students should be eager to conduct new experiments.
13. A senior scientist should not accept the new techniques suggested by a novice.
14. One scientist should not give the right to another scientist to defy his established law.
15. Positive criticism benefits the advancement of Knowledge.
16. The scientists "B" should modify his erroneous concepts if scientist "A" presents the correct and fully tested facts before him.
17. In perspective of new discoveries and inventions a scientist should be ready to change his prevalent conceptions.
18. People should be willing to change the ideas if sufficient evidences about the hollowness of their ideas are available.
19. It is impossible to defy widely held assumption in society since very long time.
20. We should not believe that small-pox, cholera and other diseases are the product of divine anger.
21. If one sneezes at the time of commencing a new task one should start it later.

22. Cooked stories by astrologers and magicians should not be preferred to scientifically based explanations.
23. For weather predictions, magicians and astrologers should not be consulted.
24. A scientist should not start his work when the way is crossed by a cat.
25. At the time of drawing inferences the scientist should draw only those conclusions which coincide with the present political ideologies.
26. Unacceptable new idea by all people should not be given due consideration.
27. Enough evidence supporting a certain idea should be provided before the idea is accepted.
28. People should read only those newspapers which are in consonance with their political ideologies.
29. The scientist should draw inferences on the basis of accurate observation.
30. If a tea company offers a bribe to any scientist, then he should not disclose the research finding about the adverse effects of tea.
31. Knowledge once accepted should not be put to test.
32. When traditional beliefs are in conflict with scientific discoveries it is better to accept the traditional beliefs.
33. Due to fast explosion of knowledge facts and theories which stand true may be disproved tommorrow.
34. Although a new theory propounded by a senior and experienced scientist "A" raised some doubts in the mind of a junior and young scientist "B". Scientist "B" should accept "A" theorty.
35. More importance should be given to the traditional beliefs than the new discoveries of science.
36. If a science teacher fails to arrive at the expected results during demonstration, he should try to discuss the possible causes of his failure.

33

Development of Scientific Attitude : An Experimental Study

M.J. Ravindranath

A Review of the literature related to science education in India would reveal that in the last few years increasing emphasis has been placed on the development of scientific attitude at the secondary school level. The literature is replete with articles describing the components of scientific attitude, and how it could be developed in students. However, one fails to notice any attempt towards construction and validation of an instrument to measure it. It was in this context that a scientific attitude test (SAS) was developed and standardised by the author along with another researcher at CASE, M.S. University of Baroda. The test was utilized as an instrument for validating a multimedia instructional strategy developed by the author for teaching biology at the secondary school level. To facilitate easy comprehension of the experiment, details have been given under three sections. Sections I gives an idea about the multimedia instructional strategy. Section II and Section III deal with the scientific attitude test and the actual experiment, respectively.

SECTION I

Multimedia Instructional Strategy

The main idea of developing a multimedia instructional strategy to teach the full course in biology at Class VIII level was to systematise the processs of science instruction. By systematisation is meant a systematic way of planning, designing, carrying out and evaluating the process of instruction through the optimal utilisation of the findings of research in human learning and communication and of men and material resources. In the study, systematisation of science instruction was effected through such steps as (i) Specification of inputs (entry behaviours) and outputs (terminal behaviours) in unambiguous and measurable terms ; (ii) analysis of the tests and content involved; (iii) identification and selection of appropriate methods and media for presenting learning experiences; (iv) provision for monitoring individual student's progress and for periodic evaluation for determining the effectiveness of the learning experiences provided; (v) pilot and field tryouts, etc. The strategy arrived at through such a procedure comprised of the following components:

1. Introduction by the teacher
2. Programmed learning material and Deviated P L M
3. Lecture method
4. Team teaching
5. Inquiry technique
6. Pupil activities and teacher dimensions
7. Discussion sessions
8. Audio-visual presentations
9. Historical background of scientists and scientific inventions
10. Summaries
11. Criterion tests and feedback
12. Exercises and assignments.

Through this strategy a complete course of eight units in biology was taught to Class VIII students of Shreyas Vidyalaya, Baroda during the year 1978-79. Development of scientific attitude being an objective attempted to be attained through the strategy, wherein different instructional components are selected and organised in a systematic manner to provide appropriate learning experiences, the potentiality of the strategy to the promotion of scientific attitude is ensured. It may be mentioned that while providing learning experiences through this strategy, deliberate efforts wherever appropriate to develop behaviours related to scientific attitude were made. A few illustrations to this effect are cited below.

1. In a view instructional units, the historical background of scientists and their contributions in terms of knowledge development in the field of science were traced.
2. Closely related to the above, instances indicating how knowledge gets refined in the light of evidence to the contrary of what exists were pointed out with a view to indicating that accumulation of knowledge is not a finished enterprise.
3. Students were encouraged to participate in group discussions. Further, they were also encouraged to carry out activities on an individual basis as well as in groups.

It should be noted that these different learning experiences with regard to their contribution towards the development of scientific attitude are to be seen in cummulative manner, and not in isolation. This is due to the view held in the investigation that the construct scientific attitude is a composite of different behaviours, and possessing any one of these would not position an individual high on scientific attitude. Precisely, in the investigation scientific attitude has been considered as a totality of different behaviours (detailed about the components of the construct have been given in Section 2).

Obviously, when concerted and deliberate attempts have been made to incorporate learnings experiences which lead to the development of behaviours related to scientific attitude, studying the extent to which it has been developed in students would reflects the effectiveness of the learning experiences provided for its development. It is in this connection that the extent of development of scientific attitude (gain from pre-test to post-test) has been considered as a criterion for validation the developed strategy.

SECTION 2

Scientific Attitude Test

As mentioned in the earlier section, in the investigation, scientific attitude has been considered as a totality of different behaviours. In the test these different behaviours have been categorized under eight components. They are :

1. Empiricism
2. Curiosity
3. Freedom from bias
4. Open-mindedness
5. Criticality
6. Intellectual Honesty
7. Seeking evidence
8. Observation

As illustrations, the behaviours enumerated under the components Empiricism and Criticality have been presented below.

Empiricism : belief in cause-effect relationship; aversion to superstition; and commitment to rationality.

Critically : analysis of various aspects of a problem or situation; reflects on observation; looks for inconsistencies in statements and conclusions; and challenges the validity of unsupported statements.

The final format of the test comprised of 24 items, with three items under each of the component. Each item in the test is a multiple choice type item with three alternatives representing least, moderate and most scientificity. The procedure for scoring being 0,1 and 2, respectively, for least, moderate and most scientificity. Cronbach's alpha coefficient for the test ranges from 0.536 to 0.764. The test has been examined for its logical validity.

SECTION 3

Actual Experiment

This section presents details about the design, the sample utilized, the experimentation, and the results.

Design

The design adopted for the experimentation was a pre-test-post-test design with two groups, viz., experimental and control groups.

Sample

The experimental and control groups considered for the study constituted of 45 students, each belonging to class VIII of one of the English-medium schools in Baroda, Viz., Shreyas Vidyalaya. The groups were matched groups, matched for their means and S.D.S. on their Class VIII science examination scores.

Experimentation

As an initial step, the scientific attitude test was administered to both the groups in the beginning of the academic year 1978-79. This served as pre-test. The experimental group was taught through the developed multimedia strategy for one complete academic year (1978-79). The control group was taught by the regular teacher of the school. Care was taken to see that the experimental groups students did not share their instructional material with the control group students. At the end of the academic year, once again the scientific attitude test was administered to both the groups as post-test.

Analysis of the Data

In all, four sets of scores were available: pre-test and post-test scores for the experimental group, and similarly, pre-test and post-test scores for the control group. Mean and S.D. of each set of scores were calculated. The student's 't' test was used for (1) finding out significant difference between means of pre-test and post-test scores of the experimental and control group separately, and (2) for comparing means of pre-test scores of the experimental group with that of the control group.

Results and Discussion

As can be seem from Table 1, both the groups show a significant difference in their means from pre-test and post-test, the difference being in favour of post-test. This indicates that there is development of scientific attitude in both the groups over the period of time (one complete academic year), since both the groups had been exposed to science teaching; of course, the experimental groups being exposed to the regular school teacher. However, the higher mean and lower S.D. obtained by the experimental group on the post-test viz-a-viz that of the control group with a relatively lower mean and higher S.D. suggest that, perhaps, learning experiences planned deliberately to develop this attribute might have contrib-

Table 33.1
't' Values for Significance of Difference Between Means of Pre-test and Post-test Scores of Experimental Group and Control Group, Separately

Groups	*No. of students*	*Nature of the scores*	*Mean*	*S.D.*	*'t' value*	*Level of significance*
Experimental group	45	Pre-test	32.26	4.94	8.47	**
		Post-test	38.02	5.61		
Control group	45	Pre-test	30.20	4.59	5.08	**
		Post-test	35.52	6.40		

** Indicates significance at 0.01 level.

uted to a greater extent for its development. To study this further, the means of pre-test and post-test scores of the experimental group have been compared with the corresponding means of the control group.

It may be observed from Table 2 that there is no significant difference between the means of the two groups on pre-test. On the contrary, on the post-test, the 't' value being 2.06 is significant at 0.05 level. The indicates that there is significant difference between the means of the two groups on post-test. The difference being in favour of the experimental group indicates that the experimental group indicates that the experimental group had develop scientific

Table 33.2
't' Values in Respect of the Comparison of the Means of the Pre-test Scores of Experimental Group with that of Control Group

Nature of scores	*Groups*	*No. of students*	*Mean*	*S.D.*	*'t' value*	*Level of significance*
Pre-test	Experimental group	45	32.26	4.94	2.0	N.S.
	Control group	45	30.20	4.59		
Post-test	Experimental group	45	38.02	5.61	2.06	*
	Control group	45	35.32	6.40		

N.S. indicates not significant ** Indicates significance at 0.05 level.

attitude to significant extent in comparison to that of the control group over a period of one academic year.

Recalling that deliberate attempts were made in the strategy to develop scientific attitude in students through incorporation of various learning experiences, the results of the comparisons made earlier suggest the possibility of planned instruction in contributing to the development of scientific attitude, and thereby reflects the effectiveness of the developed multimedia instructional strategy.

34

A Study of the Scientific Attitude and its Measurement

N.N. Shrivastava

In view of the modern development in science and its importance in today's world, science teaching has assumed a significant place in school curriculum. Along with the acquisition of knowledge and skill, application of science has also become an important aspect of science teaching in schools. It has also been realized that any amount of knowledge in science contributes little to national development and to the process of social change without developing scientific attitude among the learners. This is why development of scientific attitude through science lessons has been emphasized by science educators. Unfortunately, this important aspect of science teaching has not been studied properly by research workers in the area. The present study has, therefore, been undertaken with the following objectives in mind:

1. To develop an instrument to measure scientific attitude.
2. To compare the science teachers and non-science teachers in respect of scientific attitudes.
3. To compare science students and non-science students in respect of scientific attitudes.

Among the various scales of scientific attitude developed in foreign countries, the scale developed by Billeh and Zakhariades (1975) to measure scientific attitude was adopted in Hindi. The 36-item scale of scientific attitude covers the following six components of scientific attitude : (*a*) Rationality, (*b*) Curiosity, (*c*) Open-mindedness, (*d*) Aversion to superstitions, (*e*) Objectivity-intellectual honesty, and (*f*) Suspended judgement.

The scale-value and quartile-value of the adapted scale was verified by the method of Thurstone equal appearing interval for the selection of items. But Likert method of summated ratings was used to modify the items. Reliability of scale as calculated by equivalent forms method was 0.09 and validity as calculated by index reliability method was 0.94. Finally the scale was administered for the collection of data.

Sample

The study was confined to science and non-science teachers and science and non-science students of the educational divisional headquarters of the State of Madhya Pradesh. A random sample of 50 science teachers and 50 non-science teachers as well as 100 science and 100 non-science students was drawn from ten educational divisions of Madhya Pradesh, each division sharing equal number of cases. The data of the final sample were utilized for examining the six components of the scale by factors analysis. The factor loading after factor analysis of 6 × 6 correlation matrix provided an evidence that only two factors were dominant. A possible conclusion may be that in Indian conditions only two factors are adequate. To test the hypotheses the attitude scores were examined by analysis of variance and *t*-test methods:

Conclusions

1. All the four groups (science teacher, non-science teachers, science students and non-science students) demonstrated positive attitudes.
2. The obtained differences among the mean scores on SAS of the four groups revealed that the amount of science knowledge or general exposure to science course affected positively the scientific attitude as the value of obtained F (46.03) was significant at 0.5 level.
3. Significant difference in means of science and non-science teachers, science and non-science students also revealed that scientific knowledge helped in the formation of scientific attitude.

4. Significant difference between means of boys and girls revealed that scientific attitudes differed in respect of sex in early ages.

5. No significant difference in male and female teachers revealed that due to increase of age, sex had got no effect on scientific attitudes.

The researcher is aware of the limitations of the study. The study has been conducted on a particular population in limited sample and before generalizing its results, it should be tried on other populations and on wider samples. The researcher feels that efforts should be made to develop a new scale of scientific attitudes in Indian conditions on the basis of the results of the factor analysis.

Development of scientific attitude is one of our major objectives of teaching science in schools and as such there is also a need of situational test for measuring accurately the effectiveness of teaching science with an objective of developing scientific attitude.

35

Scientific Attitudes of In-service and Pre-service Science Teachers

D. Bhaskara Rao
B. Joseph Raju
G. Sundara Rao

Development of scientific attitude is one of the objectives of science education. The science teacher plays a major role in cultivating scientific attitudes among pupils. So this study was intended to identify the level of possession of these scientific attitudes by in-service and pre-service science teachers. The sample consisted of 36 in-service and 147 pre-service science teachers. The attitude studied are freedom from superstitions, ability to identify cause and effect relationship, and open-mindedness. It is found that the distribution of scientific attitudes under study is not normal and these are distributed independently in the samples and are independent of each other without showing any relationship. In-service teachers exhibited a little high degree of these attitudes. According to the results, 54.26 per cent are able to identify cause and effect relationship, 34.33 per cent are to superstitious and only 24.4 per cent are open-minded. This study indicates that science teachers themselves are not possessing scientific attitudes even after having a long period of science education. This study points the need for developing systematic teacher training and in-service training programmes giving stress to scientific attitudes.

In the process of identifying the goals of science education almost all the authors make a mention of scientific attitudes Dewey

(1934) cautions against the neglect of this aspect, Narendra Vaidya (1976) says that the development of scientific attitudes should not be left to chance. Allport (1935), Henry, R.E. (1964), Diedrich, Paul B. (1967), Schwab, J.J. (1962), Science Education in America (1956), Vakkins and Perry (1961), Branching (1941), Objectives of Teaching . School Science (NCERT, 1967), Bhaskara Rao (1986), Sharma, R.C. (1984), Bower, J.T. (1961), Kothari Commission (1964-66) have stressed the importance of scientific attitude and the role of the science teacher in developing these scientific attitudes.

The objectives of science teaching can be achieved only if they are taught by the teacher in the right manner. There is an awareness among science educators that if scientific attitudes are to develop from the study of science they must be taught directly and systematically in the same manner as we try to develop a mastery of scientific principles among the learners.

Though there may be a contribution from other subjects for the development of scientific attitudes, significantly a major role is to be played by science education. Imparting the science education is a responsibility of the science teacher. Hence, primarily, the teacher himself should be in possession of scientific attitude. It is logical to expect that a teacher of science or a would-be teacher of science should possess a high level of scientific attitude.

Very few research studies are available on scientific attitudes. Ira, C.D. found that well-trained, experienced teachers were possessing a high level of scientific attitudes, but, on the contrary, Bhaskara Rao, et al (1986) found that experienced science teachers were holding a low level of scientific attitudes. Reddy (1979) has also observed that scientific attitudes among student-teachers are low.

NEED OF THE STUDY

The development of scientific attitudes is recognized by educationists and curriculum planners as a major goal of science education. Among the various objectives of science teaching the objectives under the affective domain receive much lip sympathy but the least action. The teachers rarely plan their science teaching with an idea of developing scientific attitudes and almost never use any techniques of finding out whether this objective is realized or not.

In this connection, identification of the nature and quality of scientific attitudes possessed by in-service and pre-service science teachers can serve as a preliminary step because only a science teacher can provide proper atmosphere in the classroom to inculcate scientific attitude in the tiny minds who blossom into citizens.

HYPOTHESES

1. A good proportion of science graduates have to exhibit a high level of scientific attitude because of the fact that they have received sufficient science education.
2. In-service science teacher who are already working in schools should be better in having these attitude than raw graduates or pre-service science teachers who are still in the process of working themselves to become teachers.
3. There will be a high degree of association between the scientific attitudes.
4. There will be an influence on the status of scientific attitudes by the following variables—Professional Status, Sex, Qualifications, Academic Achievement, Science Discipline studied at the Graduation level, Medium of Instruction at the Graduation level, Social Status, Religion, Teaching Experience, In-service Courses attended, and Science Fairs attended.

TOOLS

The attitudes under study were Freedom from Superstitions, Ability to Identify Cause and Effect Relationship, and Open-mindedness.

For the purpose of measuring freedom from superstitions, 14 beliefs were listed in statements by the investigators and sample had to agree to disagree with the statements.

To identify the cause and effect relationship, the tool developed by Gopal Krishna, D. (1975) was adopted.

To study open-mindedness, six multiple choice test items related to it from the standardized tool of M. James Kozlow and Marshall A. Nay (1976) were used.

The final form of the test consisted of 36 items meant for

administration. The tool was administered to a try-out a sample of 50 graduates to test its reliability. The reliability of the test was found as 0.79 by using the split-half method.

SAMPLE

The tool was administered finally to a total sample of 183, out of whom 36 were-in-service and 147 were pre-service teachers.

ANALYSIS OF DATA

Attitude-wise sample distributions were computed and these distributions were tested for normality by applying the chi-square test. The degree of association between the attitudes was identified by applying the chi-square test of independence. Sample means and standard deviations were calculated for identifying the association of the variables of this study with the attitudes. The significance was tested of 0.05 level. The above analysis was attempted for studying the distribution of scientific attitudes in the sample and for attitude association and analysis of variables.

RESULTS

1. The distribution of scientific attitudes, namely freedom from superstitions, ability to identify cause and effect relationship and open-mindedness is not normal.
2. There was not much difference in the attitude between in-service science teachers and pre-service science teachers. In-service teachers exhibited a little high degree of these attitudes.
3. The results are contrary to the hypothesis that a significantly high percentage of the sample should possess a high level of scientific attitudes. The results show that 34.43 per cent (41.67 per cent in-service and 32.65 per cent pre-service) teachers were not superstitious ; 54.26 per cent of the sample (Only pre-service) were able to identify the cause and effect relationship ; and 24.04 per cent (33.33 per cent in-service and 21.77 per cent pre-service) of the sample were open-minded.
4. Quite peculiarly, none of the variables were associated with the attitudes except the medium of instruction. Among the pre-service teachers, the Telugu-medium

Table 35.1
Distribution of Scientific Attitudes

Attitude : *Freedom from Superstitions*

Sample	N	Not Superstitious	Slightly Superstitious	Superstitious	Highly Superstitious	Very Highly Superstitious	Chi-square Value
Whole sample	183	63 (34.33)	77 (42.08)	38 (20.76)	5 (2.73)	0 (0)	691.1*
In-service teachers	36	15 (41.67)	13 (36.11)	8 (22.22)	0 (0)	0 (0)	211.7*
Pre-service teachers	147	48 (32.65)	64 (43.54)	30 (20.41)	5 (3.4)	0 (0)	1010.6*

Attitude : *Ability to Identify Cause and Effect Relationship*

Sample	N	Very High Ability	High Ability	Ability	Just Able	Not Able	Chi-square value
Whole sample	183	0 (0)	9 (4.92)	36 (19.67)	36 (19.67)	102 (55.74)	1755.36*
In-service teachers	36	0 (0)	8 (22.22)	21 (58.33)	7 (19.45)	0 (0)	3.12
Pre-service teachers	147	0 (0)	1 (0.68)	15 (10.20)	29 (19.73)	102 (69.39)	2481.3 *

Attitude : *Open-mindedness*

Sample	N	Highly Open-minded	Open-minded	Not Open-minded	Chie-square Value
Whole sample	183	0 (0)	44 (24.04)	139 (75.96)	498.73*
In-service teachers	36	0 (0)	12 (33.33)	24 (66.67)	66*
Pre-service teachers	147	0 (0)	32 (21.77)	115 (78.23)	438.13*

*Significant at 0.05 level   Figures in brackets show the percentage

Table 35.2
Attitude Relationship in the Main Samples

I. *Between Freedom from Superstition and Ability to Identify Cause and Effect Relationship*

Sample	N	Chi-square Values	Df
Whole Sample	183	23.96	16
In-service teachers	36	4.96	16
Pre-service teachers	147	19.59	

II. *Between Freedom from Superstition and Open-mindedness*

Sample	N	Chi-square Values	Df
Whole Sample	183	6.83	
In-service teachers	36	7.22	8
Pre-service teachers	147	4.02	

III. *Between Freedom from Superstition and Ability to Identify Cause and Effect Relationship*

Sample	N	Chi-square Value	Df
Whole Sample	183	3.98	
In-service teachers	36	1.36	8
Pre-service teachers	147	6.56	

All chi-square values are not significant at 0.05 level
Df=16 Chi-square at 0.05 level is 26.296
Df=8 Chi-square at 0.05 level is 15.507

students were slightly open-minded than the English-medium students.

5. Another peculiarity was that scientific attitudes were distributed independently in the sample and were independent of each other.

DISCUSSION

The primary role of the science education is to develop an individual with scientific attitude in order to promote a healthy behaviour in himself and society at large. But this study shows that science teachers are not possessing a high level of scientific attitudes and the attitudes are also not distributed normally in in-service and pre-service teachers. In this context it will be necessary to incorporate the techniques of developing scientific attitudes among pupils in teacher training programmes.

The attitudes under study are also distributed independently in the sample of teachers and are independent of each other. This is a peculiar state of attitude relationship, because it is to be expected that an attitude helps in developing other attitudes. So it needs attention.

The lack of association of the variables with the attitudes under study is another significant aspect. Social attitudes and the influence of home environment may be the variables influencing the development of scientific attitudes. The nature of learning experiences and exposure to discovery situations may also be associated variables because a considerable criticism is levelled against the quality of instruction in our schools.

According to Tyler, attitudes may be developed through:

1. Assimilation from environment : for example, viewpoints held by friends and acquaintances.
2. Emotional effects of certain types of experiences, traumatic ones also included.
3. Analysis of direct intellectual process.

In the same manner, one can cultivate or develop scientific attitudes by :

1. laying emphasis on scientific reading,
2. listening to inspirational talks on science and investigation,
3. participating in discussions on various scientific issues and problems,
4. taking extensive laboratory experience into consideration,
5. understanding and solving social problems,
6. finding solutions to questions posed by students,
7. participating in science clubs, fares, exhibitions, etc.

On the whole, the conclusions of this study point towards the need for developing systematic teacher training programmes and in-service orientation programmes focussing on the scientific attitudes.

REFERENCES

1. Bhaskara Rao, D., Sundara Rao, G. and Rao, S.R.M. "Scientific Attitude of Experienced Science Teachers at Secondary School level", *The Educational Review*, XCII(4), 61-65, 1986.
2. Bhaksara Rao, Digumarti. " Development of Creativity among student". Experiments in Education XI(6), 100-103, 1983.
3. Dewey, John. "The Supreme of Intellectual Obligation, *Science Education*, February 1934.
4. Diedrich, Paul B. "Components of Scientific Attitude", *The Science Teacher*, 1967.
5. Gopal Krishna, D. A Study of Scientific Attitude and its Research to Intelligence of Graduate Students, Unpublished Master of Education Dissertation, Nagarjuna University, 1975.
6. Heiss, et al. *Modern Science Teaching*, New York : McMillan Co., 1961.
7. Henry, Richard E. "The Development of Scientific Attitudes", *The Science Teachers*, December, 1964.
8. Ira, C.D. "The Measurement of Scientific Attitudes", *Science Education*, Vol. 19, No. 3.
9. Narendra Vaidya. *The Impact of Science Teaching*, New Delhi: Oxford & IBH Publishing Co., 1976.
10. Narendra Vaidya and Rajput, J.S. (Eds.) *Reshaping Our School Science Education*, New Delhi : Oxford & IBH Publishing Co., 1977.
11. National Council for Teacher Education, *Teacher Education Curriculum : A Framework*, New Delhi : NCERT , 1982.
12. Sharma, R.C. *Modern Science Teaching*, New Delhi : Dhanpat Rai & Sons, 1984.
13. Walter, A. Thurber and Alfred, T. Collete. *Teaching Science in Today's Secondary Schools*, Boston : Allyn and Bacon Inc., 1968.

36

Scientific Attitudes of Experienced Science Teachers

Digumarti Bhaskara Rao
G. Sundara Rao
S. Raja Mohan Rao

A. INTRODUCTION

"Science is an accumulated and systematized learning in general usage restricted to natural phenomenon. The progress of Science is marked not only by an accumulation of facts, but by the emergence of the scientific method, and of the scientific attitude". An individual with right attitudes can function efficiently and an individual with scientific attitudes will bring progressive changes in the society.

Scientific attitudes can make a teacher participate properly in the social revolution. Lack of these attitudes in a teacher make him socially less useful, introvert and even indifferent. Hence, a science teacher should know and understand what the scientific attitudes are , because success in developing scientific attitudes among the pupils depends ultimately upon the teachers.

If a science teacher himself does not possess and believe in scientific attitudes, he will not be in a position to teach them. One has to remember that the children form attitudes more from examples than from abstract precept. The teacher's practice of these scientific attitudes will make a favourable and lasting impression upon the pupils.

It seems that research in necessary as the conclusions may prove to be useful in organizing special in-service programmes to teachers with the sole intention of developing the right scientific attitudes among them.

B. OBJECTIVES OF THE STUDY

1. To identify the areas of scientific attitudes where the science teachers require improvement.
2. To quantitatively identify the % of science teachers who hold different levels of scientific attitudes.
3. To identify whether the following factors have any influence on scientific attitude held by the science teachers :
   a) *Residence* : Whether the teacher is working in an urban or rural area.
   b) *Subject taught* : Whether the teacher is teaching the basic discipline studied at collegiate level, or whether teaching a science subject, irrespective of the science (biological and physical science) studied.

C. METHODOLOGY

The standardized tool of M. James Kozlow and Marshall a. Nay (1976) is employed to serve the purpose. The tool consists of 40 multiple-choice items. The areas defined in the tool are :

1. Critical mindedness
2. Suspended judgement
3. Respect for evidence
4. Honesty
5. Open-mindedness
6. Willingness to change opinion.

The tool was administered to a sample of 58 experienced (more than 10 years of service) science teachers of different high schools of Anantapur and Kurnool district of A.P. Variable-wise distribution of sample is shows in Table 36.1.

Table 35.1
Distribution of the Sample of Science Teachers

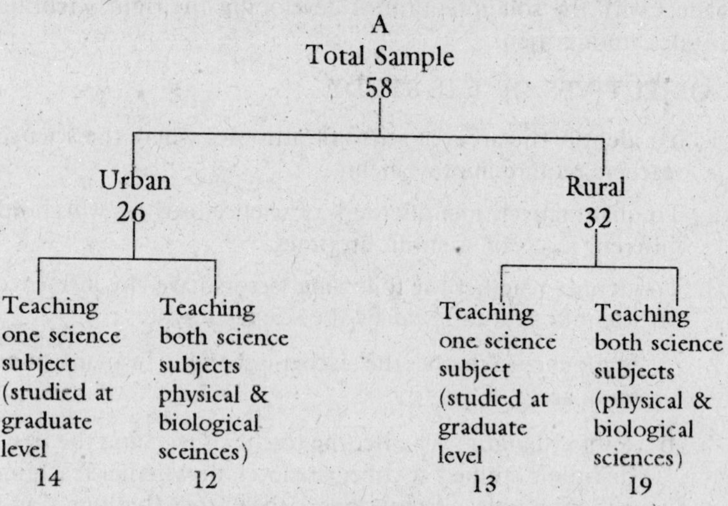

Table 36.2
Means of Scientific Attitutdes of the Sample

Variable	*N*	*Mean*	*S.D.*	*Mean difference*	*SED*	*C.R.*
Urban	26	15.5	3.42	0.25	0.987	0.25
Rural	32	15.75	4.1			
Handling one science	27	16.3	3.0	1.4	0.993	1.409
Handling both sciences	31	14.9	4.5			

For categorising the teachers into sub-groups with reference to the range of attitudes held by them, the following classification is adapted.

Table 36.3 : Range of Attitudes

Range of attitudes scores in %	*Category*
0–20	Very low
21–40	Low
41–60	Average
61–80	High
81–100	Very high

Table 36.4
Sub-Attitude (Area) Scores in Each Category

Category	*Total Attitude Scores (%)*	*Sub-Attitudes (%)*					
		I	II	III	IV	V	VI
Very low	8	27	18	30	42	49	15
Low	57	30	43	43	29	30	17
Average	35	40	25	18	23	20	43
High	Nil	3	14	9	6	1	17
Very High	Nil	Nil	Nil	Nil	Nil	Nil	8

Table 36.5 : Mean & Standard Deviations of Sub-Attitudes

Sub-Attitudes	*Group means*	*Standard deviation*
1. Critical mindedness	38.5	17.32
2. Suspended judgement	42.1	17.0
3. Respect for evidence	41.5	21.9
4. Honesty	31.5	21.0
5. Open-mindedness	26.6	2.0
6. Willing to change opinion	52.1	16.8

D. DATA ANALYSIS

The scores are converted into percentages. The necessary frequencies taken are drawn, and means and standard deviations are calculated from them. For comparison of the means of different sub-groups of the small, 'T' values are computed. The significance of the difference within each variable is tested at 0.05 level.

E. FINDINGS

1. The sample of experienced science teachers seem to be holding low scientific attitudes. 65% of the sample holds low (57%) or very low (8%) scientific attitudes. Only 35% of them hold average scientific attitudes. No one is either with high or with very high scientific attitudes.
2. The only scientific attitudes that is predominant in the sample is willingness to change opinion. At the same time, it seems that

the sample requires attention in the direction of possessing open-mindedness and honesty.

3. Location of the schools in which these experienced teachers are working does not seem to be a factor affecting the quality of scientific attitudes held by them.
4. The nature of the science discipline studied at graduation level and the subjects handled in regular class room instruction do not seem to have any influence on the nature of scientific attitudes held by them.

F. DISCUSSION

This study indicates that the quality of scientific attitudes held by this sample of science teacher is poor. Previous investigation on the scientific attitudes (Ramachandra Reddy, 1979) has also revealed that the scientific attitudes among the student-teachers are also low. It is an established dictum that the science teachers should know and understand what the scientific attitude are. The level of the attitudes held by them should be high. Science teachers should live and act their part, if they are really interested in developing scientific attitude among the pupils and in promoting the interests of the society. The seminars, symposia, in-service programmes, etc. may be useful in developing scientific attitudes among the science teachers.

G. SUGGESTIONS FOR FURTHER RESEARCH

1. The scope of this study can be extended to make a comparative study of scientific attitudes among the teachers of other sciences, social sciences and languages.
2. Studies on scientific attitudes of inexperienced and student-teachers may give better focus on this area of development of scientific attitudes.
3. Studies on the effect of economic and social back-ground, sex, personality traits on the nature of scientific attitudes held by teachers may lead to significant conclusion, which may give proper direction in cultivating the scientific attitudes.

REFERENCES

1. Bhaskara Rao, D., 'Development of Creativity among students', *Experiments in Educations*, Vol. XI, No. 6, August 1983, pp. 100-103.
2. Heiss, et. al. *Modern Science Teaching* New York : Macmillan Company, 1961.
3. Narendra Vaidya and J.S. Rajput, *Reshaping Our School Science Education*. New Delhi, Oxford & IBH Publishing Co., 1977.
4. Sharma, R.C., *Modern Science Teaching*, New Delhi : Dhanpat Rai & Sons, 1975.
5. Walter, A. Thumber and Alfred T. Collette, *Teaching Science in Today's Secondary Schools*.

37

Scientific Attitudes and Personality Traits of Prospective Science Teachers with High Pedagogic Aptitude

D. Bhaskara Rao
G. Sundara Rao
S. Aruna
L. Rathaiah

Prospective science teachers should have high pedagogic aptitude, high scientific attitudes and personality traits needed by a teacher in modern times. This study conducted on a purposive sample of B.Ed. students with high pedagogic aptitude has bearings on selection criteria to B.Ed. course and science teacher education. The trend of the data indicated that high pedagogic aptitude is related to the personality traits, radicalism and dominance. But, the low scientific attitude trend in prospective science teachers points out the need for curricular change in B.Ed. course and in-service training programmes. The authors suggest the use of composite criteria as admission requirement into teacher education. Use of pedagogic aptitude measure, past scholastic achievement and competition scholastic measure are proposed.

INTRODUCTION

Education is the main instrument for any social change and the teacher plays a vital role in any educational system. Time and again our educational committees and policies have identified the place and importance of science teacher. Unless a science teacher possesses proper personality traits and scientific attitudes, he will not be in a position to develop and nourish the personality traits and scientific attitude of his student clientele.

Many educationists have given a good number of lists of scientific attitudes. Among them, the most useful scientific attitudes are open-mindedness, critical mindedness, respect for evidence, suspended judgement, intellectual honesty, willingness to change opinion, search for truth, curiosity, rational thinking, etc.

The personality of a science teacher will definitely influence the students behaviour because, at tender age, he serves as a model to the students. The students, on many occasions, try to imitate their teachers. So, a teacher should possess right personality traits and should follow them to cultivate or develop good personality. The teacher should not be totally conservative and submissive. A mixture of radicalism with conservatism and dominance may be an useful personality traits for a teacher. He should possess a high pedagogic aptitude. For a science teacher another equally important characteristic is having scientific attitude.

Study of Ira, C.D. revealed that well trained and experienced science teachers were possessing a high level of scientific attitudes, but the study of Bhaskara Rao, et al (1986) opposed this result. Study of Reddy (1979) identified that the scientific attitudes among student teachers were low. Another study of Bhaskara Rao, et at (1988) has explained that the scientific attitudes among in-service and pre-service science teachers were low.

Darji (1975), Padma(1975), Gandhi(1977), Mehta (1977) Shah (1977), Gupta (1978), Singh (1978), Bhaskara Rao, et al(1985) have found an influence on leadership behaviour, school climate, administrative behaviour and allied educational aspects of headmasters, but not many studies are available on prospective teachers in these aspects.

NEED AND SCOPE

Aspirants to teaching profession should be individuals with high pedagogic aptitude; foundation in the subject they going to teach; having acquired the right attitudes as outcomes of the education they received and possessing personality traits suitable for teaching profession. They are the right material for teacher training.

The first thing one expects while selecting pupil teachers into Colleges of Education is that the prospective teachers should possess a high pedagogic aptitude besides possessing right attitudes and necessary personality traits. A study of scientific attitudes and personality information, and no one has tried to link personality traits with scientific attitudes of prospective science teachers. Hence this study.

Sample of this study is limited to the prospective science teachers with high pedagogic aptitude with reference to scientific attitudes and two personality traits, viz., Q_1 (**Conservatism**-- conservatism, respecting established ideas and tolerant of traditional difficulties vs. **Radicalism** -- experimenting, critical, liberal analytical and free thinking) and E (**Submissiveness**-- humble, mild, accommodating and confirming vs. **Dominance** -- assertive, independent, aggressive and stubborn).

OBJECTIVES

The objectives of this study were:

1. To identify the nature of distribution of scientific attitudes and the two personality traits, Q_1 and E.
2. To find out the degree of association between dependent variables, i.e., scientific attitudes and personality traits Q_1 and E.

HYPOTHESES

Previous science education and pedagogic aptitude are assumed to be the factors contributing to the possession of scientific attitudes and personality traits by a prospective science teacher. As the sample chosen was that of prospective science teachers with high pedagogic attitude, the hypotheses framed for verification were :

1. The distributions of scientific attitudes, personality traits Q_1 and E in this sample would not be normal and will be positively skewed, i.e. tending toward high scientific attitude, radicalism and dominance.
2. There will be association between scientific attitudes, personality traits Q_1 and E.

METHODOLOGY

TOOL : For the purpose of measuring the selected personality traits, Q_1 and E and Form A of the 16 PF Test of Raymond B. Cattell and Herbert W. Eber was used. And for measuring the scientific attitudes, the Scientific Attitudes Test of James Kozlow and Marshall A. Nay was used.

SAMPLE : The sample consisted of 68 prospective science teachers with high pedagogic aptitude scores. They were selected to the B.Ed. course into the College of Education in Guntur on the basis of an entrance examination with a predominantly pedagogic aptitude component.

Table 37.1
Discription of Sample

Variable	*Classification*	*Sample Size*
Qualification	Grudates	58
	Post graduates	10
Sex	Men	32
	Women	36
Main Subject in Degree	Biological Science	50
	Physical Science	18

ANALYSIS OF DATA

After administering and scoring of the two tools to the selected sample, the data was classified with references to the scientific attitudes and personality traits as low, average and high. To these distributions, chi-square test was applied by adopting the normal probability hypothesis and interpreted at 0.05 level of significance.

Tables 37.2 and 37.3 show chi-square values for the whole sample, Table 37.4 for sub-samples on the basis of variables — qualification, sex and discipline. 3 × 3 contingency tables were drawn and chi-square test of independence was applied to find out the degree of association between the scientific attitudes and personality traits Q_1 and E of this sample of prospective science teachers with high pedagogic aptitude.

Table 37.2
Distribution of Dependent Variables

Variable	*Low*	*Average*	*High*	*Chi-square value*	*Skewness trend of sample*
Scientific Attitudes	28	37	3	33.83*	---
Personality Factor Q_1	1	54	13	10.84*	+
Personality Factor E	6	41	21	11.90*	+

*Significant at 0.05 level + Positive trend –Negative trend
Observed frequencies are tested for normality distribution df=2 P=5.991

The sample exhibits an average an level of scientific attitudes and personality traits Q_1 and E. The distributions are not normal.

Table 37.2
Distribution of Dependent Variables

Scientific Attitude	*Low*	*Average*	*High*	*Chi-square*	*Skewness trend of sample*
1. Critical mindedness	11	50	7	4.05	+
2. Suspended judgement	17	48	3	6.25*	–
3. Respect for evidence	7	45	16	3.74	+
4. Intellectual honesty	2	63	3	19.44*	+
5. Open mindedness	45	20	3	125.50*	–
6. Willingness to change opinion	9	54	5	5.02	+

*Significant at 0.05 level + Positive trend –Negative trend
Observed frequencies are tested for normality distribution df=2 P=5.991

This sample of science teachers shows predominant by average status with reference to all the scientific attitudes. Interestingly, the sample status with reference to the scientific attitude, open mindedness appears to be low. Skewness trends in attitudes, open mindedness and suspended judgement is in negative direction.

Table 37.4 shows variable wise distribution of scientific attitudes, personality traits Q_1 and E. The variables were qualification, sex and subject (discipline) studies as main at degree level. Ten of the eighteen sub-samples are not distributed normally. The three variables qualification, sex and subject seem to be associated with the trend of distribution. Though pre-dominantly showing average status, the skewness trend of the data for these sub-samples appears to be different for scientific attitudes and personality traits Q_1 an E.

For the sub-samples of graduates, men, women teachers and biological science discipline, the skewness trend is from average towards low in scientific attitudes.

For personality traits Q_1 and E , the skewness trend in graduates, women and biological science sample is from average towards high.

As per the results, possession of scientific attitudes, was having relationship with personality traits Q_1 and E. But the two personality traits under study were not having any relationship and were distributed independently.

Table 37.5
Relationship between Dependent Variables

Dependent Variable	*Chi-square value* (test of independence)	*Conclusion*
1. Scientific attitudes with Q_1	14.59*	✓
2. Scientific attitudes with E	9.98*	✓
3. Q_1 with E	0.804	×

Df = 4 P = 9.488 * Significant at 0.05 level
✓ Denotes relationship × Denotes no relationship

FINDINGS

1. The distributions of this sample are not normal, but different skewness trend are evident for scientific attitudes and personality traits. For scientific attitudes it is negative and for personality traits it is positive. The indications of the data point out that in this sample of prospective science teachers,

   a) the scientific attitudes are low.

Table 37.4 : Variable-wise Distribution

Dependent variable	*Independent variable*	*Classification*	*Sample size*	*Frequency*			*Chi-square value*	*Skewness trend*
				Low	Average	High		
Scientific Attitudes	Qualification	Graduates	58	26	29	3	36.98*	←
		Post-graduates	10	2	8	--	1.91	
	Sex	Men	32	10	22	--	13.59*	←
		Women	36	18	15	3	30.57*	←
	Subject	Bio.Sci.	50	24	25	1	40.50*	←
		Phy.Sci.	18	4	12	2	0.66	
Personality Factor Q_1	Qualification	Graduates	58	--	46	12	10.90*	→
		Post-graduates	10	1	8	1	1.33	
	Sex	Men	32	1	26	5	3.92	
		Women	36	--	28	8	7.32*	→
	Subject	Bio.Sci.	50	--	42	8	9.88*	→
		Phy.Sci.	18	--	12	5	2.66	
Personality Factor E	Qualification	Graduates	58	7	33	18	10.66*	→
		Post-graduates	10	--	8	2	2.66	
	Sex	Men	32	2	24	6	2.18	
		Women	36	5	17	14	12.86*	→
	Subject	Bio.Sci.	50	4	31	15	8.38*	→
		Phy.Sci.	18	2	11	5	1.07	

Df = 2 P = 5.991 * Significant at 0.05 level. Observed frequency tested from normality

Ten Chi-square values are significant.

b) they are tending more towards personality traits of radicalism, i.e. experimenting, critical liberal, analytical and free thinking.

c) the tendency is towards personality traits of dominance i.e. assertive, independent, aggressive and stubborn.

2. In this sample, there is association of scientific attitude with traits, Q_1 and E . But Q_1 and E are not associated. There is a need to study their associations on a larger population to identify the degree of association.

3. In variable wise distribution, ten of the eighteen sub-samples show some distribution trends as whole sample scientific attitudes and personality traits Q_1 and E.

Though in Post graduates and Physical science samples the distribution trend appears to be towards normality, the investigators feel that the sample size is warranted to draw a conclusion.

Qualification and subject are identified as influence variables. The indications are that graduates are low in scientific attitudes, prospective teachers of biological science are low in scientific attitudes and women are low in scientific attitudes.

Women show the traits of radicalism and dominance. In the other sub-sample Q_1 and E seem to be normally distributed.

4. For the whole sample the distributions for these aspects of scientific attitudes is not normal. They are open mindedness, suspended judgement and intellectual honesty. The prospective teachers do not seem to be possessing attitude of open mindedness and suspended judgement. For the attitude of intellectual honesty the simple status in average level. The attitude critical mindedness, respect for evidence and willingness to change opinion are normally distributed.

DISCUSSION

While this is a study involving a purposive sample of prospective science teachers with high pedagogic aptitude, the investigator feel that of conclusions have some implications for deciding the selection criteria for admission of students to Colleges of Education and the process of teacher training. At this juncture, the questions that arise are :

1. Is pedagogic aptitude or useful selection criterion or would a measure of past or competitive scholastic achievement serves the criteria purpose for admitting students to B.Ed. course?
2. As the entrants to the B.Ed. course low scientific attitudes, are changes needed in the B.Ed. curriculum produce teachers with high scientific attitude?

An individual scholastic achievement may not be highly related to his personality traits. But the data of this points that high pedagogic aptitude appears to be associated with personality traits. It is commonly accepted among teacher educators that scholastic willingness to become a good teacher, personality motivation, aptitude, readiness to be a life long learner, past achievement, etc., are same of the needed qualities and hence should find a place in selection criteria. It is the opinion of the investigators that a composite criteria will be more predictive and a serve a useful purpose for selecting and admitting students to B.Ed. course. The criteria can be pedagogic aptitude, past scholastic achievement (like marks in qualifying degree) and competitive scholastic achievement of an entrance test with common components like language skills, general knowledge and subject knowledge.

The issue of low scientific attitudes among science teachers both in-service and prospective is not new. The addition in the other data is that the prospective teachers with high pedagogic aptitude are also with low scientific attitudes. This is basically an outcome of the quality of science education. What can the teacher education do to rectify this situation? While knowledge transmission can be a slow or speeded up process, attitude changes scientific attitudes in prospective teachers. Properly learnt scientific knowledge can help in acquiring scientific attitudes. If the teacher educator has to take up this role, the duration of B.Ed. course should be of two years or more on the pattern of an Integrated Degree like in Regional Colleges of Education. In this context, a study of the scientific attitudes of prospective teachers undergoing, integrated course may yield useful data. There is also a need for developing well planned in-service programmes for developing attitudes among science teachers.

Some more questions arising out of this study are, is this phenomenon of low scientific attitude restricted to students of Biology discipline? Are the women teachers more radical and domi nant than men? This area of scientific attitude, personality traits and pedagogic aptitude needs to be investigated more in detail and depth. Extending the study to other and larger samples will definitely yield useful data.

REFERENCES

Atkinson, Rita L. et al, *Introduction to Psychology*, (8Ed), New York, Hercourt Brace Jovanorvich, Inc., 1983.

Bhaskara Rao, D., "Development of Creativity among Student", *Experiments in Education*, XI(6), 100-103, 1983.

Bhaskara Rao, D., Parisudha Rao, N. and Sundara Rao, G., "Personality Characteristics of Educational Administrators", *Journal of the Institute of Educational Research*, 9(3), 8-13, November, 1985.

Bhaskara Rao, D., Sundara Rao, G. and Mohana Rao, S.R., "Scientific Attitudes of Experienced Science Teachers at Secondary School Level', *The Educational Review*, XCIII(4), 61-65, April 1986.

Bhaskara Rao, D., Sundara Rao, G. and Rathaiah, L., "The Science Teacher has a definite role", *The Hindu*, September 27, 1988, p.19.

Garrent, H.E., *Statistics in Psychology and Education*, Bombay, : Allied Pacific Pvt., Ltd., 1961.

Pritam Singh, "Scientific Attitude - Its Development and Assessment", in Vaidya, N. and Rajput, J.S. (Eds), *Reshaping our School Science Education*, New Delhi : Oxford an IBH Publishing Co., 1977.

Vaidya, Narendra, *The Impact Science Teaching*, New Delhi : Oxford & IBH Publishing Co., 1976.

38

Scientific Attitude in Secondary School Students

Digumarti Bhaskara Rao

Every citizen of the present modern world sees the countless manifestations of science all around him. There is no aspect of man's life today which has not been influenced by science one way or the other. This is because we live in an age of scientific culture. Science has shrunk the world and totally changed the human outlook. In fact, science now has all pervading influence on every sphere of human activity. Further, modern science is no longer confined to the surface of this globe, its sphere of achievement reached beyond the earth.

There has been, recent times, rapid addition of knowledge to the world of science. Great achievements of science and technology and the use of these scientific achievements in promoting the well-being of mankind through their application in the field of industry, communication, transport, engineering, agriculture and medicine have made science more important than ever before. Science has radically transformed the material environment of the citizens of the modern world; and, of course, it has its significant role in promoting culture and spiritualism either directly or indirectly.

Teaching of everyday science for everybody has become an avoidable part of general education. Nobody questions its inclusion as a subject in the school curriculum. It is included in a school's curriculum for the same reasons as any other subject, but in addition, science inculcates certain special values peculiar to it and which not other subject can provide. Besides satisfying the usual needs for its inclusion as a subject in the curriculum - such as intellectual, cultural, moral aesthetic, utilitarian as well as vocational values - science learning provides training in scientific method, and also helps to develop a scientific attitude of mind and scientific aptitude in the learner. Therefore, science is now a compulsory subject in every system of school education right from the elementary level.

Since the beginning of the twentieth century, science educators have included the development of scientific attitude among the general aims of science education. The scientific attitude, by its very name, tend, to be associated solely with the area of science. There are many components in scientific attitude; and those components of scientific attitude were explained by Curtis (1924), Arthur (1935), Davis (1935), Noll (1935), Caldwell and Lundeen (1936), Ebel (1938), Lampkin (1938), Wesseli (1941), Caldwell and Curtis (1943), Heiss, Obourn and Hoffman (1950), Rethinking Science Education (1960), Baumel and Berger (1965), Diederich (1967), Young and Schmid (1968), Smith, Krouse and Atkinson (1969), Vaidya (1976), Kozlow and Nay (1976), Okey, (1982), Gauld (1982), Schibeci (1983), Sharma (1984), Bhaskara Rao (1989), Bhaskara Rao, et al. (1989) and so on. Though many have explained various components of scientific attitude, there is a general agreement among these educationists that a person who has scientific attitude - is open-minded; shows intellectual honesty; suspends judgement until he gets accurate information; looks for cause and effect relationship; listen to other's point of view; is free from superstitions; has a habit of basing judgement on fact; is willing to change opinions on the basis of evidence; is curious concerning things; is desirous for experimental verification; and is with rational thinking.

For many science educators, the importance of the scientific attitude is so obvious that no arguments is required to support its inclusion among those things which a school science course should aim to develop in pupils. The idea is also reinforced by the fact that there has been little, if any, argument against its inclusion among the aims of science education. However, while it may be obvious that the scientific attitude is important in the professional lives of scientific and that students learning about science should also become aware of the motive power which impels scientists in their work, it is not a simple matter to move on to the conclusion that school students, many of whom do not intend to become scientists, should actually be encouraged to adopt this attitude themselves.

DESIGN OF THE STUDY

Design is the heart of any research study. Due care and importance were given in formulating objectives and hypotheses, and in selecting variables, sampling techniques, sample and tools.

Considering their role in determining scientific attitude in the secondary school pupils, variable such as sex, type of school management, location of the school, type of schools, and language used in instruction were selected for the present study.

Objectives were identified keeping the different aspects of the present study. There are: 1. To find out the scientific attitude possessed by the secondary school pupils and 2. To compare the scientific attitude of boys and girls, private and government school pupils, urban and rural school pupils, residential and non-residential school pupils, and English and Telugu medium school pupils.

Hypotheses were formulated taking the above objectives into consideration. The hypotheses are: 1. Secondary school pupils will possess high scientific attitude, and 2. There is no significant difference between the level of scientific attitude possessed by boys and girls; private and government school pupils; urban and rural school pupils; residential and non-residential school pupils; and English and Telugu medium school pupils.

Stratified Sampling Technique was found to be the most appropriate technique for the present study because the present study involves splitting of the sample into a good number of groups according to various variables. Random sampling technique was also employed to select the pupils from each group.

Regarding the **size of the sample**, 600 was found to be appropriate. This was found suitable because the study involves due intensity and detail, a sample with more than 600 pupils would involve a lot of resources, and the more important one, the time; and less than 60 pupils would also bring about problems of representativeness.

Only the pupils studying in tenth class in secondary school were included in the sample. This decision was taken because the children's attitudes get formed intensively at the age of 14 and/or 15.

The research tool occupies a major role in any study because it is useful in the collection and analysis of data to draw conclusions. As the present study is a deliberate and intensive one, the standardized 'Scientific Attitude Scale' of J.S. Sood and R.P. Sandhya was used to study the scientific attitude of secondary school pupils.

ANALYSIS OF DATA

Analysis of data is studying of the tabulated material in order to determine the inherent facts or meaning which involves breaking down the complex factors into simpler units and putting these units together in new arrangements for purpose of interpretation.

The total score of the scientific attitude of each was taken into consideration to find out the level of scientific attitude possessed by the each sub-sample as well as total sample of the study. The maximum score that a pupil can get is 180 and the minimum is 36. In the present study, the highest score secured by a pupil was 161 and the lowest was 87.

For the purpose of classification of the level of the scientific attitude possessed by the sample, the scientific attitude level was categorized by using normal probability of distribution. The pupil who score 14 and below was kept in low scientific attitude group, who scored between 115 and 149 was put in average scientific attitude group, and who scored 150 and above was placed in high scientific attitude group.

The mean scores were used to identify the level of scientific attitude possessed by the pupils and to compare the sub-sample variation. The values of standard deviation were used to measure the spread or dispersion of scores in a distribution. The critical ratios were calculated to test the significant difference in the means of two sub-samples of each variable.

The Tables 38.1 and 38.2 give the analysis and values of statistical procedures used in the study.

It is clear, from the Table 38.1, that the pupils studying in

Table 38.1 : Level of Scientific Attitude possessed by the Secondary School Pupils

Variable	*Sample Size*	*Mean*	*Standard deviation*	*Mean difference*	*S.Ed*	*Critical Ratio*
Whole	600	131.53	17.97			
Boys	300	129.29	23.67	1.45	1.69	0.85#
Girls	300	130.74	18.39			
Private	300	133.32	15.17	4.09	1.49	2.74*
Government	300	129.23	20.82			
Urban	300	127.8	17.97	7.53	1.39	5.41*
Rural	300	135.33	16.07			
Residential	120	140.68	15.98	12.19	1.71	7.12*
Non-Residential	480	128.49	19.42			
English Medium	120	135.5	13.01	4.91	1.45	3.38*
Telugu Medium	480	130.59	18.22			

Not significant at 0.01 level * Significant at 0.01 level

secondary schools hold an average level of scientific attitude. As per the standard deviation, there is a little bit high dispersion of scores in the units of the samples. The mean difference increases from sex to type of school through type of school management, medium of instruction and location of the school. As per the values of the critical ratio, there is significant difference in the level of scientific attitude possessed by all the variables except variable sex.

Further, the chi-square (X^2) test was applied to test the divergence of observed results of variables from those expected on the hypothesis equal probability (Null hypothesis). The chi-square values were calculated for each sub-sample and are given in Table 38.2.

Table 38.2 : Distribution of Scientific Attitude in the Secondary School Pupils

Variable	*Sample Size*		*Low*	*Average*	*High*	X^2
Whole	600	f_o	86	419	96	1.35#
		f_e	96	408	96	
Boys	300	f_o	43	212	45	1.02#
		f_e	48	204	48	
Girls	300	f_o	43	207	50	0.65#
		f_e	48	204	48	
Private	300	f_o	16	244	40	30.5*
		f_e	48	204	48	
Government	300	f_o	70	175	55	15.2*
		f_e	48	204	48	
Urban	300	f_o	53	208	39	2.28#
		f_e	48	204	48	
Rural	300	f_o	33	211	56	6.26#
		f_e	48	204	48	
Residential	120	f_o	8	73	39	15.74*
		f_e	19.2	81.6	19.2	
Non-Residential	480	f_o	78	346	56	6.83#
		f_e	76.8	326.4	76.8	
English Medium	120	f_o	5	86	29	15.74*
		f_e	19.2	81.6	19.2	
Telugu Medium	480	f_o	81	333	66	1.88#
		f_e	76.8	326.4	76.8	

Not significant 0.01 level *Significant at 0.01 level

f_o = frequency of occurrence of observed or experimentally determined facts

f_e = frequency of occurrence expected on some hypothesis.

Scientific attitude is distributed normally, as the chi-square values are not significant, in the whole sample of secondary school pupils and also in boys and girls, urban and rural pupils, non-residential school pupils and Telugu medium school pupils.

Scientific attitude is not distributed normally in pupils studying in private and government schools, residential schools in English medium schools.

CONCLUSIONS AND DISCUSSION

The present study has resulted in drawing the following conclusions which may be utilized in improving the present state of affairs in the school science education and science teaching.

The **scientific attitude in secondary school pupils is average. The distribution of this scientific attitude in the tenth class pupils is normal**. Davis, Caldwell and Lundeen, Gopal Krishna and Keurst found that the pupils under study were superstitious. Here, our problem is why this present sample holds an average level of scientific attitude and how it can be developed and promoted more than the existing level.

The facilities like library, laboratory, audio-visual aids, exposure to eminent personalities, participation in fairs, exhibitions, etc., will help in the inculcation and promotion of scientific attitude in the individuals. The above mentioned facilities are not so abundantly available in our schools, particularly in government schools. Another significant feature is the possession of scientific attitude by the teachers who teach it have influence directly or indirectly in the classrooms. A teacher without proper scientific attitude cannot develop or promote it. The studies of Bhaskara Rao, Sundara Rao and Mohan Rao and Bhaskara Rao, et al. have identified the experienced and prospective science teachers held only low scientific attitude. The above mentioned factors may be the reasons for the average level of scientific attitude possessed by the secondary school pupils.

Now, it is the right time for the identification of necessary factors for the promotion of scientific attitude in the school children. Some of the factors necessary for the promotion of scientific attitude are: Informative experiences about the attitude object, situations arising in solving a problem, pleasant emotional experiences, well equipped science labs, group decision making, encouragement in the cultivation of desirable attitude; engaging in wide reading in general science; preservation of democratic procedures, suggesting problems that need to collect evidence to form conclusions, stressing the need for adequate data before arriving at conclusions; through assimilation of environment, through the emotional effects, through traumatic experiences, through direct intellectual processes, providing opportunity for the analysis of problem; amount of scientific knowledge or exposure to general science courses; work experience, providing proper laboratory facilities, taking the pupils to fairs, exhibitions, excursions, field trips, zoos, parks, industries, natural, habitats of plants and animals, through direct teaching of the required scientific attitude; following the pupils to mingle with various peer and intellectual groups; exposing them to the eminent personalities like scientists, social reformers, etc. All the science educators must try to promote the scientific attitude in the pupils by implementing the above mentioned factors which are feasible in their own educational set up.

The scientific attitude in both boys and girls is average and there is no difference in the level of scientific attitude possessed by them. The distribution of scientific attitude in these two groups is also normal. As the sample is from the co-educational schools, except A.P. Residential Schools, this result states that if the opportunities are equal to either sex, they can complete with each other equally in any area. This is contrary to the result of the study made by Shrivastava. The science teacher must try to promote scientific attitude in the pupils through the above suggested procedures.

The scientific attitude possessed by the pupils in private and government schools fall under average category. The pupils of private schools hold a little bit high scientific attitude than those of government schools. But both the sub-samples are with an average scientific attitude. The distribution of it in both of them is also not normal. It is tending towards high scientific attitude category in private schools, where as it is reverse in government schools. Many people say that the facilities available in the private schools are good. The quality of teaching will also be good as there are better facilities. Another important thing for this quality is that inferior teaching will be questioned immediately without any delay, which is not possible in the case of government schools. The teachers will teach throughout a pupil's school career in private schools as the teacher works in the same school for a long time without any transfers, and also as he understands the flaws and potentialities of his pupils. All these factors will play a significant role in promoting scientific attitude, which may be adopted in government school also.

The scientific attitude possessed by the pupils of urban and rural schools is average. The rural pupils hold a high level of scientific attitude than urban pupils, but the distribution in both the sub-samples is normal. On the contrary, earlier studies of Bhaskara Rao, Sudara Rao and Mohan Rao and Gopal Krishna found that the locale did not influence the possession of scientific attitude. The result of this study is a surprising one, because, the urban schools are supposed to be equipped well with all facilities, and the qualities of teaching may also be good as many people say. This study indicates that if conducive facilities are providing to the rural pupils they will score as better as urban pupils.

The scientific attitude possessed by the pupils of residential and non-residential school is average. But when compared, the pupils of residential schools are with very high scientific attitude than those of non-residential schools. The distribution of scientific attitude is normal in the pupils of non-residential schools, but its trend is towards high level in the pupils of residential schools. The facilities that are available in A.P. Residential Schools are no where available either in private school or in other government schools. The facilities that available, the teaching learning schedule they follow, the intelligence of the pupils of residential schools might have helped in possessing such a high scientific attitude. So, these facilities may be extended to other types of schools.

The scientific attitude in the pupils of the English and Telugu medium schools is also average. The English medium pupils hold a bit of high scientific attitude than the Telugu medium pupils. The distribution of scientific attitude is normal only in Telugu medium pupils, but its concentration is tending towards high level in English medium pupils. Normally, if there exists any language barrier, the pupils of Telugu medium should possess better scientific attitude than those of English medium as the former understand the subject very easily as Telugu is their mother tongue. The finding of this study supports that the better facilities available in a school facilitate the development of scientific attitude.

It is worthwhile to consolidate the results regarding the influence of variables on scientific attitude. **The pupils studying in private schools, rural schools, English medium schools and residential schools hold relatively better scientific attitude than their counter parts. There is no influence of sex on scientific attitude. All the pupils of five variables hold an average scientific attitude.** The significant influence of the above variables makes it clear that the facilities that are available, and the conducive teaching learning atmosphere are the deciding factors in cultivating and promoting the scientific attitude.

SUGGESTIONS FOR FURTHER RESEARCH

The present study brings to light a good number of new areas to be studied by the future researchers. The areas and variables which are not covered by this study may be put to test to enlighten the factors associated with the inculcation and development of scientific attitude. The researcher may think of the following areas to study in detail.

1. Studies on scientific attitude may be extended to other educational levels, viz. primary, college and university levels.
2. Studies may be taken up to find out the effect the independent variables on dependent variables in the cases of controlled and experimental groups as this study has not used any special groups.
3. Studies may be conducted on the influence of habits, hobbies, etc., in the promotion of scientific attitude.
4. Studies may be taken up to find out the effect of environmental and psychological factors on the inculcation and development of scientific attitude.
5. Studies about the scientific attitude possessed by the teaching community may be taken up this factor has a great role to play in the development of scientific attitude in the classrooms.
6. Studies may be carried out on the use of audio-visual teaching aids, teaching-learning strategies, laboratory and library facili-

ties as these have greater influence on the inculcation and promotion of scientific attitude.

7. Studies on the role of science exhibition, science clubs, science museums and other scientific centers in developing scientific attitude may be carried out.

BIBLIOGRAPHY

Baumel, Howard, B. and J. Joel Berger, "An Attempt to measure Scientific Attitudes". *Science Education* 49 (April 1965), 267-269.

Bhaskara Roa, D., "Objectives of Science". *Science Promoter* 2 (October 1989), 701-703.

Bhaskara Rao, Digumarti, "Utilization of Community Resources in Science Teaching". *Junior Scientist* 23 (February 1986), 5-6.

Bhaskara Rao, Digumarti, "Science Education in Secondary Schools: A Reflection". *National Conference on New Educational Policy: Its Need and Concept.* Department of Education, Hindu College, Moradabad, India, October 10-12, 1987.

Bhaskara Rao, .Digumarti, "Natural Science Instruction in India: An Analysis". *Second International Congress for Research on Activity Theory.* University of Helsinki, Lahti Research and Training Centre, Helsinki, Finland, May 21-25, 1990.

Bhaskara Rao, Digumarti, "Biological Basis of Learning". *Second International Conference on Differentiated Psychology of Learning - Its Fundamentals and Application.* Martin Luther University, Halle-Wittenberg, Germany, September 16-18, 1991.

Bhaskara Rao, Digumarti. *A Comparative Study of Scientific Attitude, Scientific Aptitude and Achievement in Biology at Secondary Schools Level.* Unpublished Ph. D. thesis, Osmania University, Hyderabad, 1989.

Bhaskara Rao, Digumarti. *Audio Visual Teaching Aids*, Guntur: Nagarjuna Publishers, 1989.

Bhaskara, Rao, Digumarti. *Teaching of Biology.* Guntur: Nagarjuna Publishers, 1991.

Bhaskara Rao, D. G.S. and S.R.M. Rao, "Scientific Attitudes of Experienced Science Teachers at Secondary School Level". *The Educational Review* XCII (April 1986), 60-65.

Bhaskara Rao, D., *et. al*., "Scientific Attitudes and Personality Traits of Prospective Science Teachers". *Progressive Educational Herald* 3 (January 1989), 62-66.

Buch, M.B. Chief editor. *Third Survey of Research in Education*. New Delhi: National Council of Educational Research and Training, 1987.

Caldwell, Otis W. and Gerhard E. Lundeen, "Students Attitudes Unfounded Beliefs". *Science Education* 15 (May 1931), 246-266.

Davis, Ira C., "The Measurement of Scientific Attitudes". *Science Education* 19 (October 1935), 117-122.

Diedrich, Paul B., "Components of the Scientific Attitude". The *Science Teacher* 34 (February 1967), 23-24.

Ebel, Robert L., "What is the Scientific attitude?" *Science Education* 22 (January 1938), 1–5.

Gauld C.F. and A.A. Haukins, "Scientific Attitudes: A Review". *Studies in Science Education* 7 (1980), 129–161.

Haney, Richard E., "The Development of Scientific Attitude". *The Science Teacher* 31 (December 1984), 33–35.

Heiss, E.D., E.S. Obourna and C.W. Hoffman. *Modern Science Teaching*. New York: The MacMillan Co., 1950.

Keurst, Arthur, J. Ket, "The Acceptance of Superstitious Beliefs among Secondary Schools Pupils". *Journal of Educational Research* 32 (May 1939), 673-685.

Krishna, D. Gopala. *A Study of Scientific Attitude and its Relation to Intelligence of Graduate Students*. Unpublished M.Ed. dissertation, Andhra University, Watair, 1975.

Kozlov, James A. and Marshall. A. Nay, "An approach to Measuring Scientific Attitudes". *Science Education* 60 (April-June 1976), 147-172.

Lampkin Jr. Richard H., "Scientific Attitude". *Science Education* 22 (December 1938), 353-357.

Noll, Victor H., "Measuring the Scientific Attitude". *Journal of Abnormal and Social Psychology* 30 (July 1935), 145-14.

Okey, James, R., "The Scientific Attitude and Science Education: A Critical Reappraisal". *Science Education* 66 (1982), 109-121.

Pearl, Richard E., "The Present Status of Science Attitude Measurement: History, Theory and Availability of Measurement Instruments".

School Science and Mathematics LXXIV (May-June 1974), 375-381.

Ravindranath, M., "Development of Scientific Attitude: An Experimental Study". *Journal of Indian Education* (November 1983), 28-32.

Schibeci, R.A., "Selecting Appropriate Attitudinal Objectives for School Science". *Science Education* 67 (October 1983), 595–603.

Shrivastava, N.N., "A Study of the Scientific Attitude and its Measurement". *Indian Educational Review* (January 1983), 95-97.

Vaidya, Narendra. *The Impact Science Teaching*. New Delhi. Oxford & IBH Publishing Co., 1976.

39

A Comparative Study of Scientific Attitude, Scientific Aptitude and Achievement in Biology at Secondary School Level

Digumarti Bhaskara Rao

Teaching of science has become an unavoidable part of general education. Nobody questions its inclusion as a subject in the school curriculum. It is included in a school's curriculum for he same reasons as any other subject, but in addition, science inculcates certain special values peculiar to it and which no other subject can provide. Besides satisfying the usual needs for its inclusion as a subject in the curriculum—such as intellectual, cultural, moral, aesthetic, utilitarian as well as vocational values - science learning provides training in scientific method, and also helps to develop a scientific attitude of mind and scientific aptitude in the learner. Therefore, science is now a compulsory subject in every system of school education right from the elementary level.

Since the beginning of the twentieth century, science educators have included the development of scientific attitude among the general aims of science education. The **scientific attitude**, by its very name, tends to be associated solely with the area of science. There are many components in scientific attitude; and those components of scientific attitude were explained by Curtis (1942), Arthur (1935), Davis (1935), Noll (1935), Caldwell and Lundeen (1936), Ebel (1938), Lampkin (1938), Wesseli (1941), Caldwell and Curtis

(1943), Heiss, Obourn and Hoffman (1950), Rethinking Science Education (1960), Baumel and berger (1965), Diederich (1967), Young and Schmid (1968), Smith, Krouse and Atkinson (1969), Vaidya (1976), Kozlow and Nay (1976), Okey (1982), Gauld (1982), Schibeci (1983), Sharma (1984), Bhaskara Rao (1989), Bhaskara Rao et al. (1989), and so on. There is a general agreement among these people that a person who has scientific attitude (1) is open-minded, (2) shows intellectual honesty, (3) suspends Judgements until he gets accurate information, (4) looks for cause and effect relationship, (5) listens to other's point of view, (6) is free from superstitions, (7) has a habit of basic judgement on fact, (8) is willing to change opinions on the basis of evidence, (9) is able to distinguish between fact and theory, (10) is curious concerning things, (11) is desirous for experimental verification, and (12) is with rational thinking.

Scientific aptitude is a complex of interacting hereditary and environmental determinants producing predispositions or abilities in science. It is a potentiality for future accomplishment in science without regard to past training and achievement. It appears to be dependent upon a variety of factors such as study skills, motivation, persistence in learning a subject, socio-economic factors, cultural background, interests attitudes.

Biology is one of the compulsory subjects in the school curriculum in Andhra Pradesh. It paves way to the career deciding course at +2 stage. The acquisition of the knowledge of biological terms, principles and concepts, a clear understanding of them, the ability to use such knowledge in different situations in life and in the development of skills should be the outcomes of teaching and learning of biological science. Moreover, the pupils should develop a proper attitude towards the study of biology and an active interest in the subject, besides appreciating the importance of biology in human life and civilization. It also helps in improve their ability and capacities in science. There is a need for awareness in biology for the full pledged development of a child.

RESEARCH DESIGN

Objectives

Objectives are identified, keeping the different aspects of the present study, in view. The main objectives of the study are:

1. To find out the scientific attitude and scientific aptitude possessed by the secondary school pupils along with their achievement in biology.
2. To find out the association among scientific attitude, scientific aptitude, and achievement in biology of secondary school pupils.
3. To compare and scientific attitude, scientific aptitude, and achievement in biology of boys and girls, rural versus urban schools, English versus Telugu medium schools, private versus government schools and residential versus non-residential schools.

Variables

Considering their role in determining scientific attitude, scientific aptitude and achievement in biology, **variables** such as boys versus girls, rural versus urban schools, English versus Telugu medium schools, private versus government schools, and residential versus non-residential schools are selected.

Hypotheses

Hypotheses are formulated taking the above objectives into consideration. These hypotheses are formulated only in positive manner, but the sub-hypotheses are formulated in Null form. The main hypotheses of the present study are:

1. Secondary school pupils will possess high scientific attitude.
2. Secondary school pupils will be good at biology achievement.
3. There will be a significant positive association among scientific attitude, scientific aptitude and achievement in biology.

Sampling Technique

Stratified sampling technique, after making a detailed study of different techniques of sampling, is found to be the most appropriate technique for the present study. This technique is found to be the most suitable one because the present study involves splitting of the sample into a good number of groups according to different variables. Through stratified sampling only it is possible to divide the sample into different groups or strata and choose pupils from each of these groups. Random sampling technique is also employed to selected pupils from each group.

Sample

Regarding the **size of the sample,** 600 is found to be appropriate. This is found suitable because the study involves due intensity and detail, a sample with more than 600 pupils would involve a lot of resources, and the more important one, time. Less than 600 pupils would also bring problems of representativeness. Hence, 600 is considered to be appropriate number for the sample.

Only the **pupils studying in tenth class** in secondary schools of Guntur district, Andhra Pradesh are included in the sample. This decision is taken because the children's attitudes and aptitudes get formed intensively at the age of 14 or 15. The biology of tenth class decided their career at +2 stage.

A **sample of 600 tenth class pupils** is taken into consideration. Out of the total sample, 300 pupils are from rural schools and 300 pupils are from urban schools. All the pupils selected from rural schools are of Telugu medium as thee are no English medium school in rural area. Regarding of type of school, 120 pupils are from residential schools and 180 are from non-residential schools. As regards the management of schools, 210 pupils are from government schools and 90 are from private schools in rural area. An equal representation is given to both boys and girls in the rural sample. As regards the schools in urban area, 180 pupils are selected from Telugu medium schools and 120 selected from English medium schools. Out of these 300 pupils, 90 are from government schools and 210 are from private schools. In this urban area also, equal proportion is given to both boys and girls. With the above splitting of the total sample into various strata, the final sub-group sample sizes are; boys -300 and girls -300, rural -300 and urban -300, private schools -300 and government schools -300, Telugu medium -480 and English medium -120, and residential schools -120 and non-residential schools -480. Thus, the total sample (600) is split into different groups.

Tools

The tools occupy a major role in any research study because they are useful in the collection and analysis of data to draw the conclusions. Construction and standardization of good tool itself is a major research work. As the present study is a deliberate and intensive one, the available standardized tools are employed. Scientific Attitude Scale of J.K. Sood and R.P. Sandhya, and Kerala University Science Aptitude Test of Nair, et al. are used to study scientific attitude and scientific aptitude of secondary school pupils respectively. As these two tools are too large, and time consuming ones, the marks in biology scored in the pre-public examination of the tenth class of the district are taken into consideration to assess the achievement of pupils of biology.

CONCLUSION AND DISCUSSION

The three areas, viz., scientific attitude, scientific aptitude and achievement in biology will play a major role in moulding a child's character and in selecting a career deciding course in science. So, every pupil of school education should posses good scientific attitude and scientific aptitude and should secure good scores in biology. The present study in these three major areas with their association with each other has resulted in drawing the following conclusions which may be utilized in improving the present state of affairs in the schools science education.

1. The scientific attitude in secondary school pupils is average. The distribution of this scientific attitude in tenth class pupils is normal.

Davis[1], Caldwell and Lunden[2], Gopal Krishna[3], and Keurst[4] found that the pupils were superstitious. Here, our problem is why this sample holds an average level of scientific attitude and how it can be developed and promoted more than the existing level.

The facilities like library, laboratory, audio-visual aids, exposure to eminent personalities, participation in fairs, exhibition, etc. will help in the inculcation and promotion of scientific attitude in the individuals. The above mentioned facilities are not so abundantly available in our schools, particularly in government schools. Another significant feature is the possession of scientific attitude by the teachers who teach it have influence directly or indirectly in the classrooms. A teacher without proper scientific attitude cannot develop or promote it. The studies of Bhaskara Rao, Sundara Rao and Mohan Rao[5] and Bhaskara Rao, *et al*[6] have identified the experienced and prospective science teachers held low scientific attitude respectively. The above mentioned factors may be the reasons for average level of scientific attitude possessed by the secondary school pupils.

Now, it is the right time for the identification of the necessary factors for the promotion of scientific attitude in the school children. Some of the factors necessary for the promotion of scientific are: Informative experiences about the attitude object, situations arising in solving a problem, pleasant emotional experiences, well-equipped science labs, group decision making, encouragement in the cultivation of desirable attitude[7]; engaging in wide reading in general science[8]; preservation of democratic procedures, suggesting problems that need to collect evidences to form conclusions[9]; through assimilation of environment, through the emotional effects, through traumatic experiences, through direct intellectual processes, providing opportunity for the analysis of problem;[10] amount of scientific knowledge or exposure to general science course;[11] work experience;[12] providing proper laboratory facilities; taking the pupils to fairs, exhibitions, excursions, field trips, zoos, parks, industries natural habitats of plants and animals; through direct teaching of the required scientific attitude;[13] allowing the pupils to mingle with various peer and intellectual groups; exposing them to the eminent personalties like scientists, social reforms, etc. All the science educators must try to promote the scientific attitude in the pupils by implementing the above mentioned factors that are feasible in their own educational set up.

2. The scientific attitude in both boys and girls is average and there is no difference in the level of scientific attitude possessed by them. The distribution of scientific attitude in these two groups is also normal.

As the sample is form the co-educational schools, except A. P. Residential School, this result states that if the opportunities are equal to either sex, they can compete with each other equally in any area. This is in contrary to the result of the study made by Shrivastava.[14] The science teacher must try to promote scientific attitude in the pupils through the above suggested procedures.

3. The scientific attitude possessed by the pupils in private and government schools fall under average category. The pupils of private school hold a little bit high scientific attitude than those of

government schools. But both the sub-samples are with an average scientific attitude. The distribution of it in both of them is also not normal. It is tending towards high scientific attitude category in private schools, where as it is reverse in government schools.

Many people say that the facilities available in the private schools are good. The quality of teaching will also be good as there are better facilities. Another important thing for this quality is that inferior teaching will be questioned immediately without any delay, which is not possible in the case of government schools. The teachers will teach throughout a pupils schools career in private schools as the teacher works in the same school for a long time without any transfers, and as he understands the flaws and potentialities of his pupils. All these factors will play a significant role in promoting scientific attitude, which may be adopted in government school also.

4. The scientific attitude possessed by the pupils of urban and rural schools is average. The rural pupils hold a high level of scientific attitude than urban pupils, but the distribution in both the sub-samples is normal.

One the contrary, earlier studies of Bhaskara Rao, Sundara Rao and Mohan Rao[15] and Gopal Krishna[16] found that the locale did not influence the possession of scientific attitude. The result of this study is a surprising one, because the urban schools are supposed to be equipped well with all facilities, and the quality of teaching may also be good as many people say. This study indicates that if conducive facilities are provided to the rural pupils they will score as better as urban pupils.

5. The scientific attitude in the pupils in English and Telugu medium schools is average. The English medium pupils hold a bit of high scientific attitude than the Telugu medium pupils. The distribution of scientific attitude is normal only in Telugu medium pupils, but its concentration is tending towards high level in English medium pupils.

Normally, if there exists any language barrier, the pupils of Telugu medium should possess better scientific attitude than those of English medium as the former understand the subject very easily as Telugu is their mother tongue. The finding of this study supports that the better facilities available in a school facilitate the development of scientific attitude.

6. The scientific attitude possessed by the pupils of residential and non-residential schools is average. But, when compared, the pupils of residential schools are with very high in scientific attitude than those of non-residential schools. The distribution of scientific attitude is normal in the pupils of non-residential schools, but its trend is towards high level in the pupils of residential school.

The facilities that are available in A.P. Residential School are no where available either in private schools or in other government schools. The facilities that are available, the teaching learning schedule they follow, the intelligence of the pupils of residential schools might have helped in possessing such a high scientific attitude. So, these facilities may be extended to other types of schools.

7. The scientific attitude in the secondary school pupils is average. The distribution of it in the whole sample is also normal.

In a study, Jose[17] found that 70% of 9th class pupils were with average scientific aptitude. As scientific aptitude is a potentiality for future achievement in scientific endeavour and as it is a complex of interacting hereditary and environmental determinants producing such potentialities or predispositions for future accomplishment, it is necessary to develop scientific aptitude among the pupils. Development of scientific aptitude is dependent on a variety of factors. The presence of certain study skills; persistence in learning and motivation; satisfaction derived from learning a subject, evaluation procedures that are followed in education; cultural background; socio-economic factors, interests; attitudes are some of the important factors that promote of scientific aptitude.

8. The scientific aptitude in boys and girls is average and the distribution of it is also normal in both the sub-samples.

The teachers must try to promote and the pupils must try to attain high scientific aptitude.

9. The scientific aptitude is average in the pupils of private and government schools. The pupils of private schools possess a bit of high scientific aptitude than those of government schools. The distribution in the two sub-samples is not normal. The scientific

aptitude concentration is towards high in private schools and vice versa in government schools.

The managements of government schools should provide the facilities conducive to the promotion of scientific aptitude, and the teachers must utilize available resources in a best possible manner to promote scientific aptitude in the pupils.

10. The scientific aptitude in the pupils of urban and rural school is average, but the urban pupils possess a little bit high scientific aptitude than rural pupils. The distribution of scientific aptitude in both the sample is normal.

The facilities that are available are the causes to be said for this difference in the scientific aptitude possessed by the urban rural pupils. The authorities concerned should try to promote the educational facilities in rural schools.

11. The scientific aptitude is average the pupils of Telugu and English medium schools. The pupils of English medium possess a little bit high scientific aptitude than those of Telugu medium. The scientific aptitude trend is towards high scientific aptitude in English medium pupils, and its trend is towards low category in Telugu medium pupils.

This results states the language plays a role in the development of scientific aptitude. But, this result may be because of the facilities that facilitate it in the English medium schools.

12. The scientific aptitude is average in residential and non-residential schools. But the pupils of residential schools are superior to those of non-residential schools. But the pupils of residential schools are superior to those of non-residential schools. The trait distribution is normal in both the cases.

The educational facilities that prevail in residential schools should be provided in other types of schools.

13. The achievement in biology is average in the whole sample. The distribution of achievement in the sample is also normal.

Biology is one of the compulsory subjects at 10+ stage and paves way to the career deciding branching at +2 stage, and ultimately to professional course like medicine, agriculture, veterinary, science etc. So, there must be good achievement in biology at secondary school level. The biology achievement depends on the aptitude, attitude and interest of the pupils towards biology, conducive teaching learning atmosphere, laboratory and library facilities that are available, the efforts put in by the pupil, use of audio-visual aids in teaching learning situations, etc. If these facilities are adequate, then there will be good achievement, otherwise the result would be nullified.

14. The achievement in biology in average in both boys and girls. The distribution of biology achievement is also normal in boys and girls of secondary schools.

This states there is no influence of sex on achievement. But, according to the Second International Science Study,[18] girls scored higher boys in biology in some countries. And the same study concludes that the differences between boys and girls in sciences achievement were greater in the physical science than in the life sciences. But contrary to this, the findings of Thakur[19] and Second International Science Study,[20] indicate that boys were superior to girls in science achievement. The result of this study and the contradictory findings of the above mentioned studies indicate that if proper facilities are provided, both boys and girls will score well in any subject.

15. The achievement in biology in the pupils of private and government schools is average. There is no significant difference in the achievement of these two groups, but the achievement is not normally distributed in both the cases. The achievement concentration is more or average, but slightly towards low in private schools. It is towards high in government schools.

The pupils who exposed to better library, laboratory, audio-visual and teaching facilities in private schools scored better than the counterparts. These facilities should be provided in government schools also.

16. The biology achievement is average in the pupils studying in urban and rural schools. The achievement of the rural pupils is better than the urban pupils. The achievement trend in rural schools is more towards high achievement, whereas it is towards low achievement in urban pupils.

This states that the locate of the school is no barrier if there is better learning atmosphere.

17. The biology achievement in the pupils of Telugu and English medium schools is average. The English medium pupils are slightly better than Telugu medium pupils. The achievement distribution in Telugu medium pupils is normal, but it is concentrated in average level of achievement in English medium pupils.

As all the English medium schools selected for the present study are in urban areas, they are equipped well in many aspects. This might have been the cause for better achievement in English medium pupils. So, the Telugu medium schools should be equipped on par with their counter parts.

18. The biology achievement is average in the pupils of residential and non-residential schools. The pupils of residential schools are very superior to those of non-residential schools. The distribution of achievement is very high towards high achievement in the case of residential schools. The achievement trend is towards low in the case of non-residential schools.

As earlier discussed, the residential schools are really contributing a lot in the area of education to the society. It will be nice if this type of status is provided to all other types of schools.

19. The association among scientific attitude, scientific aptitude and biology achievement is highly significant and positive. In the whole sample, there is high association in between scientific attitude and biology achievement when compared with the two other combinations, and the lesser association is seen in between scientific attitude and scientific aptitude.

It is a good sign and in support of the previous studies. Pillai,[21] Jose,[22] Sujatha,[23] Thampya[24] and Nair and Joseph[25] found the there was a significant positive association between biology achievement and scientific aptitude. In contrary to the previous studies, great association is seen in between scientific attitude and biology achivement than scientific aptitude and biology achievement. This study reveals that the above three factors are inter-related. So, the teachers must try to cultivate and promote them in association of these three aspects. If we develop scientific attitude or scientific aptitude, this in turn leads to the development of the other and these who will help in bringing better achievement.

20. The association among scientific attitude, scientific aptitude and biology achievement in both boys and girls is highly significant and positive. The association between scientific attitude and biology achievement is more and the association between scientific aptitude and biology achievement is relatively less in boys. The same trend is seen in the case of girls too.

These results state that there is no influence of sex on the association of these aspects, but on the contrary, studies of Ganguly,[26] Nair and Joseph,[27] and Skaria[28] revealed that scientific aptitude was highly associated with the academic success in science achievement of girls. This study gives an indication that if the opportunities are provided equally to either sex, they can compete equally in all aspects.

21. The association among scientific attitude, scientific aptitude and biology achievement in the pupils studying in private and urban schools is highly significant and positive. When the associations are compared with each other, the association in private school is relatively more than in government schools. There is a great association between scientific attitude and scientific aptitude in private schools when compared with the two other combination, and a lesser association is seen in between scientific aptitude and biology achievement. In the case of government schools, much association is seen in between scientific attitude and biology achievement and lesser association is same as in private schools.

These results indicate that where there are better educational facilities there is better association in between the traits. So, one should try to improve the physical facilities of school as well as teaching efficiencies.

22. There is a high significant and positive association among scientific attitude, scientific aptitude and biology achievement in the pupils of urban and rural schools. Relatively more association is seen in urban schools than in rural schools. In urban schools, much association is seen in between scientific attitude and biology achievement and the less association is seen in between scientific aptitude and biology achievement. Much association is in between scientific attitude and scientific aptitude, and the less is in between scientific aptitude and biology achievement in rural schools.

The urban pupils are supposed to be exposed to more knowledge and better facilities. This may be the reason for better association among traits in urban pupils than in rural pupils. The rural pupils may be exposed to science fairs, exhibitions, zoos, parks, industries, laboratory experiments, books, etc., for better accomplishment.

23. The association among scientific attitude, scientific aptitude and biology achievement is highly significant and positive. Relatively there is much more association in Telugu medium pupils than in the English medium pupils. In both the cases, the higher association is between scientific attitude and biology achievement and the lesser one between scientific attitude and scientific aptitude.

There will be a language hindrance in understanding a phenomenon in English medium pupils, whereas it is a mother tongue to the Telugu medium pupils. To the English medium pupils, the schools and home environments are entirely different. So, there may be equal possession of the psychological traits and achievement in biology in Telugu medium pupils.

24. There is significant and positive association among scientific attitude scientific aptitude and biology achievement in the pupils of residential and non-residential schools. A very high association is seen in non-residential schools than in residential schools. In non-residential schools, much association is seen in between scientific attitude and scientific aptitude and the lesser one between scientific attitude and biology achievement. In residential schools, relatively much association is seen between scientific attitude and biology achievement and the lesser one between scientific attitude and biology achievement. Surprisingly there is no association between scientific aptitude and biology achievement in residential schools, and these two are independent of each other.

As there is association among the three factors, the promotion of them must be taken into consideration. It will be worthwhile if the reasons for the independent nature of scientific aptitude and biology achivement are identified.

On the whole, the psychological traits, scientific attitude and scientific aptitude, and achievement in biology are average in the samples. There is highly significant and positive association among scientific attitude, scientific aptitude and biology achievement.

The science educators must try to promote the level of scientific attitude and scientific aptitude possessed by the samples, and should try to improve the achievement in biology. If necessary steps are taken, our pupils will accomplish and achieve any thing in science education.

SUGGESTIONS FOR FURTHER RESEARCH

The present study bring to light a good number of new areas to be studied by the future researchers. The areas and variables which are not covered by this study may be put to test to enlighten the factors associated with the inculcation and development of scientific attitude and scientific aptitude, and the other factors associated with the achievement in biology. So, the researchers may think of the following areas to study in detail.

1. Studies on scientific attitude, scientific aptitude and achievement in biology may be extended to the other educational levels, viz., primary and college levels at district as well as state level.
2. Studies on scientific attitude, scientific aptitude and achievement in other subjects may be taken up.
3. Studies may be conducted on scientific attitude, scientific aptitude and achievement either independently or combindly at various levels of educations, areas and variables.
4. Studies may be taken up to find out the effect of independent variables on dependent variables in the cases of controlled and experimental groups as this study has not used any special controlled variables.
5. Studies may be conducted to find out the effect of environmental and psychological factors on the inculcation and development of scientific attitude and scientific aptitude.
6. Studies about the scientific attitude and scientific aptitude possessed by the teaching community may be taken up as this factors has a great role to play in the development of scientific attitude and scientific aptitude in class rooms.
7. Studies can be taken up to identify the reasons for the possession of average scientific attitude and scientific aptitude and average biology achievement as found in this study.

8. Studies may be carried out to identify the factors effecting the lesser levels of possession of scientific attitude, scientific aptitude and achievement in biology in the case of variables studies in this study.
9. Studies may be conducted on the use of audio-visual teaching aids, laboratory and library facilities available in the schools as these have greater influence on the inculcation and development of scientific attitude and scientific aptitude and achivement in biology and other subjects.
10. Studies are required on the role of exhibitions, science clubs, science museums and other scientific centers in developing scientific attitude, scientific aptitude and achivement in biology.

References

1. Ira C. Davis, "The Measurement of Scientific Attitudes", *Science Education* 19 (October 1935), 117-122.
2. O.W. Caldwell and Gerehard E. Lundeen, "Students' Attitudes regarding Unfounded Beliefs", *Science Education* 15 (May 1931), 246-266.
3. D.C. Krishna, *A Study of Scientific Attitude and Its Relation to Intelligence of Graduate Students* (Master of Education Dissertation, Andhra University, 1975).
4. Arthur, J. Ker Keurst, "The Acceptance of Superstitious Beliefs among Secondary School Pupils", *Journal of Education Research* 32 (May 1939), 673-685.
5. D. Bhaskara Rao, G. Sundara Rao and S.R. Mohan Rao, "Scientific Attitudes of Experienced Science Teachers at Secondary School Level", *The Educational Review* XCII (April 1986), 60-65.
6. D. Bhaskara Rao, G. Sundara Rao, A. Aruna and L., Rathaiah, "Scientific Attitudes and Personality Traits of Prospective Science Teachers", *Progressive Educational Herald* 3 (January 1989), 62-66.
7. Richard E. Haney, "The Development of Scientific Attitude", *The Science Teacher* 31 (December 1984), 33-35.
8. Eldwood D. Heiss, Ellsworth S. Obourn and Charles W. Hoffman, *Modern Science Teaching* (New York : the MacMillan Co., 1950), 129 -130.

9. *Ibid.*

10. Tayler, As cited by Eldwood D. Heiss, Ellsworth S. Obourn and Charless W. Hoffman, *Op. cit.*, 24-25.

11. N.N. Srivastava, "A Study of the Scientific Attitude and its Measurement", 1975. As cited By M.B. Buch, chief ed., *Third Survey of Research in Education* (New Delhi: National Council of Educational Research and Training. 1987), 507.

12. B.G. Kulkarni, "An Investigation into the Attitudes of Pupils, Parents and Teachers towards Work Experience", 1975. As cited in *Third Survey of Research in Education, op. cit.*, 543.

13. Eldwood D. Heiss, Ellsworth S. Obourn and Charless W. Hoffman, *op. cit.*

14. N.N. Srivastava, "A Study of the Scientific Attitude and Its Measurement", *Indian Journal of Education* (January 1983), 95-97.

15. D. Bhaskara Rao, G. Sundara Rao, S.R. Mohan Rao, *op. cit.*

16. D. Gopal Krishna, *op. cit.*

17. Jose K.M., *A Comparative Study of the Biology Achievement of High-Average- and Low-Science Aptitude of Secondary School Pupils* (Master of Education Thesis, University of Calicut, 1987).

18. Willward, J. Jacobson and Rodney L. Doran, *Science Achievement in the United States and Sixteen Countries : A Report to the public*, Second IEA Science Study (The International Association for the Evaluation of Education Research, Teachers College, Columbia University, New York, 1988), 59-66.

19. R.S. Thakur, "A Study of the Scholastic Achievement of Secondary School Pupils in Bihar". As cited by M.B. Buch, ed., *Second Survey of Research in Education* (Baroda : Society for Ednl. Research and Development, 1979). 362.

20. Willard J. Jocobson and Rodney L. Doran, *op. cit.*

21. Kamala S. Pillai, "The Relative Efficiency of Science Aptitude and Intelligence to predict Biology Acheivement", *Experiment in Education* XIV (December 1986), 171-175.

22. Jose K.M. *op. cit.*

23. Kumari B. Sujatha, *The Relative Efficiency of Science Aptitude, Science Interest and Attitude towards Science in predicting Biology Achievement of Secondary School Pupils* (Master of Education, Thesis, University of Calicut, Calicut, 1987).

24. M.P. Thampy, *A Study of the Interaction of Science Aptitude and Attitude towards Science on Biology Achievement of Secondary*

School Pupils (Master of Thesis, University of Calicut, Calicut, 1984).

25. A.S. Nair and S. Joseph, *An Experimental Study of the Overlap of Intelligence and Science Aptitude with Educational Outcomes in Biology measured using Host's Taxonomy* (Department of Education, University of Kerala, 1978).

26. D. Ganguly, D. Ghose, S. Chatterji and M. Mukherji, "An Investigation into the Validity of a Scientific Knowledge and Aptitude Test". As cited in *Third Survey of Research in Education, op. cit.* 485.

27. A.S. Nair and S. Jeseph, *op. cit.*

28. S. Skaria, A study of the Attainment of Essential, Concept in Biology in Relation to science Aptitude to Secondary School Pupils (Master of Education Thesis, University of Calicut, 1984).

40

Brief Summary of Literature on Scientific Attitude

Digumarti Harshitha
Digumarti Pushpa Latha

Science educators have recognized that scientific attitudes are the most important outcomes which would result from science teaching. Much experimentation has been carried on the in the field of measuring attitudes and opinions, and most of them have come from the sociologists and the social psychologists. Science teachers and educators are not, however, unaware of the need of some valid and reliable research on the measurement of scientific attitude, its influence on certain school achievements and its development. There is much literature on scientific attitude, but it is concerned only with defining and developing it, not on measuring it in various groups of people. With this it is clear that little was done about the measurement of scientific attitude, particularly at school level.

Shrivastava[1] in his study found that science teachers, non-science teachers, science students, and non-science students, demonstrated positive scientific attitude.

Bhaskara Rao, Sundara Rao and Mohan Rao,[2] on the contrary, in their study with experienced secondary school science teachers found that 65 per cent of the sample hold low scientific attitudes. Only 35 per cent of them hold average scientific attitudes, and unfortunately no one was with high scientific attitudes.

In another study Bhaskara Rao, et al.[3] found the prospective science teachers were also holding low scientific attitudes. The only scientific attitude that was predominant in experienced science teachers was willingness to change opinion to a greater extent. The gradually decreased scientific attitudes were suspended judgement, respect for evidence, critical mindedness, honesty, and open mindedness.[4]

Davis[5] found that the high school pupils in Wisconsin were not superstitious. The high school pupils of Wisconsin seemed to have a fairly clear concept of the cause and effect relationship, but they did not seem to be able to recognize the adequacy of a supposed cause to produce the given result.

Caldwell and Lundeen[6] stated that high school seniors believed slightly more than 20 per cent of a list of superstitions. The high school seniors were apparently affected by about 22 per cent of the superstitious ideas which they were familiar.

In supporting to the study of Caldwell and Lundeen, Gopal Krishana[7] found that the college degree students were not free from superstitions.

Beliefs in superstitions does not decline with the advancement in grade levels despite the fact that the majority of 500 pupils belonging to 7,8 and 9th grades were presumably taking or had taken science courses.[8]

Downing[9] when tested the established conclusion, i.e. 'any increased power in critical processes or in scientific attitudes develops independently of, or possibly despite, science instruction', found that there was a fairly uniform and gradual increases in abilities from grade, viz., 2500 pupils in grades 8 through 12. Ravindranath[10] also came to similar conclusion that there was development of scientific attitude in both controlled and experimental pupils of Class VII over a period of one academic year. The experimental group had developed scientific attitude to a considerable degree in comparison with the controlled group.

Shrivastava[11] in his study obtained the result that the knowledge of science or general exposure to science course affected the scientific attitudes positively.

According to Gopal Krishna,[12] the graduate students who completely devoted to the science subject only have scientific attitude. So far as specialization was concerned, the students of Zoology, Botany and Chemistry had definite advantage over the other students regarding the development of scientific attitude.

Shrivastava[13] also found that scientific knowledge helped in the formation of scientific attitude.

Contrary to the studies of Shrivastava and Gopal Krishna Downing[14] found that the students who had not studied science scored average and high scores on the test of the scientific thinking that those who studies science. Gopal Krishna[15] concluded that scientific attitude is not the sole monopoly of science subjects, but is equally capable of being developed among students of non-science subjects also. There was no evidence that science subjects, as they are conveniently taught, have higher powers of scientific thinking.[16] Alpern[17] also reports that there was no significant relation between the science courses a student had taken and his ability to select sound procedures to test hypotheses. The high school students have not developed the skill as a result of their instruction. Another study of Baumel and Berger[18] revealed that the students who scored high scientific attitude were not necessarily those with high grades in science and the students who scored low were not necessarily those with low grades in science.

Shrivastava,[19] based of his study, stated that scientific attitudes differed in respect of sex in early ages, but no significant difference in male and female teachers revealed and this was due to advancement in age. This was in support of the result of Gopal Krishna.[20]

Bhaskara Rao, Sundara Rao and Mohan Rao[21] and Gopal Krishna[22] found that location, rural and urban, did not influence the possession of scientific attitude.

Science educators have realized the importance of inculcating and developing scientific attitudes among different groups of pupils. Development of scientific attitudes can be achieved only through many direction and associated behavioural factors.

Baumel and Berger[23] hypothesize that the scientific attitudes may be developed with the help of the following factors : (1) Scientific attitudes may of acquired by students at all ability; (2) The science teachers needs to evaluate not only the knowledge achievement of the students but also their growth in scientific attitude; (3) The student with scientific attitudes will more effectively cope with problem in school and community; (4) Success in developing scientific attitudes dependents ultimately on the teacher. The teacher through his actions must be able to convince the students that scientific attitudes are an integral part of his behaviour. His intellectual honesty, willingness to amidt error, listening to others' ideas, and dealing, with facts in an unbiased law make a favourable and lasting impression upon pupils.

The teachers, according to Swarnamma,[24] failed to develop scientific attitude among the pupils of upper primary classes. This was supported by the study of Davis[25] which stated that the teachers of Wisconsin did not consciously attempt to develop the characteristics of scientific attitude. If pupils have acquired these characteristics, they have acquired them by some process of thinking or experiences outside the classroom. So they must try to develop scientific attitudes.

Heiss, Obourn and Hoffman[26] suggest the following methods for the development of scientific attitudes among school pupils : (1) Preservation of democratic procedures; (2) Suggesting projects which give the pupils experience in problem solving; (3) Suggesting problems that require the collecting of evidence of forming conclusion; (4) Stressing frequently the need for adequate data before arriving at a conclusion and that conclusions based on insufficient data should be accepted tentatively; and (5) Dealing with misconceptions most effectively by creating attitudes that will serve as checks on accepting half-truths and superstitions.

Tayler[27] has summarized the findings of the learning studies which reveal the ways in attitudes are developed. They include: (1) Through assimilation from the environment. The things that are assumed by the people round about us, the point of view that is held by our friends and acquittances: (2) Through the emotional effects of certain kinds of experiences. In general, if one has had satisfying experiences in a particular connection, he develops an attitude favourable to some content or aspect of that experiences; (3) Through traumatic experiences, i.e. experiences that have a deep emotional effect; and (4) Through direct intellectual process. In some instances when we see the implications of particular behaviour, when we analyze the nature of a particular object or process, we are led to develop an attitude favourable or unfavourable to it from the knowledge which we gain from this intellectual analysis.

Tayler further proposes suggestions for planning learning experiences to build desirable attitudes such as (1) Increase the degree of consistency of the environment; (2) Increase the opportunities for making satisfying adjustments to attitude formation; and (3) Provide opportunity for the analysis of problem situations so that a pupil may understand and then test intellectually in the desirable attitude.

Study of Srivastava[28] reveals that the amount of scientific knowledge or general exposure to science courses had impact on scientific attitudes positively, and scientific knowledge helped in the formation of scientific attitudes.

Kulakarni[29] found that the work experience was effective in inculcating in the pupils love of scientific attitude.

Studies of Curtis, Blair and Goodson, and Vicklund seem to show that direct teaching does modify the attitudes of young people, and the study made of Curtis gave rather clear evidence that pupils who engage in wide reading in general science develop scientific attitudes more than those who study only single subject.[30]

The scientific attitudes which have been explored are attributes of intellectually and emotionally mature individuals persons who not only behave outwardly in desirable ways but also understand why they act as they do.

There are implications in what has been said for the education of teachers as well as for the instruction of school children. It has been said that 'you can't teach something you don't know'. A corollary to this generalization might be this: 'pupils cannot learn attitudes that their teachers do not have. It may very well be that the first step in meeting this challenge to science education will consist of an inward look upon for own knowledge and value systems. Science teachers have a responsibility. It is to them that the public turns for an understanding of science not just the facts of science, or the skills, but also for a perspective that relates science to all other areas of human experience.

References

1. N.N. Shrivastava, "A Study of the Scientific Attitude and its Measurement", *Indian Education Review* (January 1983), 95–97.

2. D. Bhaskara Rao, G. Sundara Rao and S. Raja Mohan Rao, "Scientific Attitudes of Experienced Science Teachers at Secondary School Level", *The Educational Review* XCII (April 1986), 61–65.

3. D. Bhaskara Rao, et al.,"Science Attitudes and Personality Traits of Prospective Science Teachers", *Progressive Educational Herald* 3 (Jan. '89), 62–66.

4. D. Bhaskara Rao, G. Sundara Rao and S.R. Mohan Rao, *op. cit.*

5. Ira C. Davis, "The Measurement of Scientific Attitudes", *Science Education* 19 (October 1935), 117–122.

6. O. W. Caldwell and Gerhard E. Lundeen, "Students' Attitudes regarding Unfounded Beliefs", *Science Education* 15 (May 1931), 246–266.

7. D. Krishna, A Study of Scientific Attitude and its Relation to Intelligence of Graduate Students (Unpublished Master of Education Dissertation, Andhra University, Waltair, 1975), 196 pp.

8. Arthur J. Ker Keurst, "The Acceptance of Superstitious Beliefs among Secondary School Pupils", *Journal of Education Research* 32 (May 1939), 673–685.

9. Elliot R. Downing, "Some Results of a Test on Scientific Thinking", *Science Education* 20 (October 1936), 121.

10. M. J. Ravindranath, "Development of Scientific Attitude : An Experimental Study", *Journal of Indian Education* (Nov. 1983), 28–32.

11. N. N. Shrivastava, *op. cit.*

12. D. Gopal Krishna, *op. cit.*

13. N. N. Shrivastava, *op. cit.*

14. Elliot R. Downing, *op. cit.*

15. D. Gopal Krishna, *op. cit.*

16. Elliot R. Downing, *op. cit.*

17. Morris L. Alpern, "The Ability to Test Hypotheses", *Science Education* 30 (october 1946), 220–229.

18. Howard B. Baumel and J. Joel Berger, "An Attempt to Measure Scientific Attitudes", *Science Education* 49 (April, 1965), 267–269.

19. N. N. Shrivastava, *op. cit.*

20. D. Gopal Krishna, *op. cit.*

21. D. Bhaskara Rao, G. Sundara Rao and S. Raja Mohan Rao, *loc. cit.*

22. D. Gopal Krishna, *op. cit.*

23. Howard b. Baumel and J. Joel Berger, "An Attempt to Measure Scientific Attitudes", *Science Education* 49 (April 1965), 267–269.

24. G. Swarnamma, "An Enquiry into the Teaching of Biology in the upper Primary Schools of Kerala", Ph.D. Education, Kerala University, 1978. As cited by M. B. Buch, Chief. ed., *Third Survey of Research in Education* (New Delhi : NCERT, 1987), pp. 568–569.

25. Ira C. Davis, *loc. cit.*

26. Eldwood D. Heiss, Ellsworth S. Obourn and Charles W. Hoffman, *Modern Science Teaching*. New York : The Macmillan Co., 1950.

27. Tayler. As cited by Eldwood D. Heiss, Ellswroth S. Obourn and Charles W. Hoffman, *Ibid.*

28. N.N. Srivastava, "A Study of the Scientific Attitude and its Measurement", Ph.D. Education, Patna University, 1975. As cited by M.B. Buch : Third Survey of Research in Education, *op. cit.*, p. 507.

29. B. G. Kulakarni, "An Investigation into Attitudes of Pupils, Parents and Teachers towards Work Experience", 1975. As cited by M. B. Buch, *Third Survey of Research in Education*, Ibid., p. 543.

30. Eldwood D. Heiss, Ellsworth S. Obourn and Charles W. Hoffman, *op. cit.*, pp. 129–130.